3D ORIGAMI
ART

3D ORIGAMI ART

Jun Mitani

CRC Press
Taylor & Francis Group
Boca Raton London New York

CRC Press is an imprint of the
Taylor & Francis Group, an **informa** business

AN A K PETERS BOOK

CRC Press
Taylor & Francis Group
6000 Broken Sound Parkway NW, Suite 300
Boca Raton, FL 33487-2742

Printed on acid-free paper
Version Date: 20160411

International Standard Book Number-13: 978-1-4987-6534-3 (Paperback)

Library of Congress Cataloging-in-Publication Data

Names: Mitani, Jun, 1975- author.
Title: 3D origami art / Jun Mitani.
Description: Boca Raton : CRC Press, 2016. | Includes index.
Identifiers: LCCN 2015050125 | ISBN 9781498765343 (acid-free paper)
Subjects: LCSH: Origami. | Three-dimensional modeling.
Classification: LCC TT872.5 .M565 2016 | DDC 736/.982--dc23
LC record available at https://lccn.loc.gov/2015050125

Visit the Taylor & Francis Web site at
http://www.taylorandfrancis.com

and the CRC Press Web site at
http://www.crcpress.com

Contents

Preface

Paper, one of the ordinary, essential items found around us, is used not only as canvas for drawing and painting but also as a fun material for the creative activity of making shapes through cutting, folding, and gluing. It seems that by cutting and gluing paper, any shape can be made. But if your action is limited to folding, you may feel that only limited shapes can be made. But is that really true? Now take a look into the world of origami, making shapes only through folding. You will find a magical, fascinating space of geometry woven with a variety of representations.

With a centuries-old history in Japan, origami play to fold paper into different shapes is familiar among people regardless of age. Now there are fans of origami all over the world. In its long history, origami techniques of forming a sheet into intended shapes have progressed. The background of this is an accumulation of mathematical knowledge about origami to establish theories on origami design. Furthermore, the advent of computer programs to perform the necessary computations for origami design moved the world of origami forward dramatically.

Many of you may have played by folding paper in your childhood. Origami in the twenty-first century has evolved amazingly from what it was before. In particular, the folding action along a curve and forms having curved surfaces are part of a new origami field realized by the use of computers.

This book introduces three-dimensional creations derived from computation and explains the design method. Many of them have a three-dimensional structure composed of curved surfaces, and some have complicated forms. But the background theory underlying the creations in this book is very simple. Surely you will be surprised to find that totally different-looking origami forms are designed from a common theory. This book contains many photos and design drawings called *crease patterns*. All of these crease patterns are available on my webpage http://mitani.cs.tsukuba.ac.jp/book/3d_origami_art/, so everyone can download and print them.

But I forgot one important thing. The origami creations in this book do not use square paper but use various shapes of sheets such as rectangles and regular polygons. Sometimes gluing may be necessary to stabilize the final shape. A little relaxing of rules greatly widens the range of shapes you can create.

It is my great pleasure to present information in this book so that you can enjoy the mystery and acquire mastery of three-dimensional origami, making lovely, geometric, three-dimensional structures out of a set of lines and curves drawn on a two-dimensional plane.

Prologue: Origami Basics

Origami is one of our familiar playtime activities, but its origins are not at all clear. This introduction covers the historical background of origami, its geometric nature, and the essential information for designing origami.

P.1 Road to Modern Origami

It is thought that origami play of folding cranes and boats began to spread in Japan in the early Edo period (the seventeenth century), although the period depends on how origami is defined. Before that, a decorative origami culture existed in Japan that included wrapping paper (*tatou*) and paper attached to gifts (*noshi*). Origami is deeply rooted in Japanese tradition. Today origami is used worldwide as a term indicating paper-folding play. However, paper folding is an old, established practice in Europe as well. So, the idea that "origami originated from Japan and then spread all over the world" is not entirely correct.

A collection of variations of traditional Japanese origami cranes was published in 1797, more than 200 years ago, entitled *Hiden Senbazuru Orikata* [*Secret Crane Folding Patterns*]. The book focuses on connected cranes and covers as many as 49 types of completion drawings together with information on how to cut a sheet. Many of you may think origami uses a square sheet and should not be cut, but the book tells us that in the past, a more freewheeling style was used to create origami works. *Hiden Senbazuru Orikata* is said to be the world's oldest origami play book. But I learned while writing this book that a text had been discovered in Kuwana, Mie, which dates back several years before *Hiden Senbazuru Orikata*. Origami history has many facts that are yet to be revealed. Paper-folding play has a centuries-old history in Japan and has been loved by people of all generations.

Original modern origami was probably introduced after the middle of the twentieth century. Akira Yoshizawa's contribution to origami is highly valued both at home and abroad. It was he who refined origami into an art and established the method of describing the folding procedure using figures and symbols. Various new folding techniques have been developed since about 1980, when the concept of "design" was introduced to origami. Recent years have seen structurally complex works, such as insects. These are far-removed from classical origami in sophistication. Complicated origami with many folds is called *complex origami*. The First International Meeting of Origami Science and Technology was held in 1989, advancing the study from a scholarly point of view. In 2014, the sixth meeting was held in Tokyo with about 300 participants from 30 countries, and as many as 140 of the latest origami study results were presented.

After 2000, with the spread of computers, many software programs that support origami design and simulate how sheets of paper transform with folds were introduced. Computer-aided origami design and studies are now also known as *computational origami*.

Another new, active movement is the application of origami techniques in engineering fields—the folding of satellite solar panels, the folding of car airbags, and robotics, seeking more use in various industries.

P.2 Origami and Crease Pattern

The first step to getting into origami is to look carefully at the relation between the folded "shape" and the "fold lines" on the unfolded sheet. There are two fold line types—the "mountain" (or "ridge") fold and the "valley" fold. The graphic showing the layout of these fold lines is called a *crease pattern* (Figure P.1). Different from the crease pattern, the *diagram* explains the folding process using figures and symbols.

When you unfold an actual origami work, you may find certain lines left unfolded at the time of completion (but were folded as a mark during the folding process). These lines are called *auxiliary lines* and are generally not included in the crease pattern.

In indicating crease patterns, mountain folds are often colored in red and valley folds in blue. In a monochrome setting, mountain folds are generally shown by dash-dot or solid lines and valley folds by dashed lines (in this book, mountains are shown by solid lines and valleys by dashed lines for visibility).

P.3 Theory of Flat Folding

Most of the traditional origami works are folded flatwise during the folding process. Even an origami crane wing is folded flatwise all the way through the final step. Folding a sheet flatwise is a basic origami action and is called *flat folding*.

Fold lines resulting from flat folding are always straight. So, a flat-folded origami work has a crease pattern with a set of straight lines. Figure P.2 is a finished "bird" from flat folding and its crease pattern. You may see the crease pattern represented by a set of straight lines.

Many studies have been done on the flat-folded state because it is the general origami state. Focusing on the point where fold lines meet, the following two principles

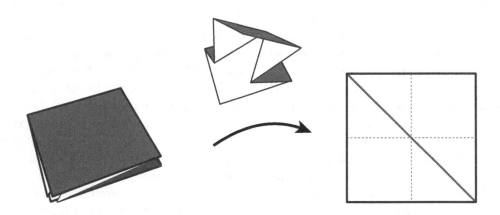

Figure P.1   Unfolded sheet has a crease pattern with mountain and valley lines.

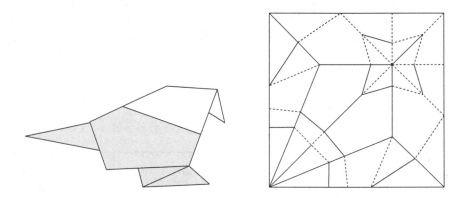

Figure P.2   "Bird" and its crease pattern (viewed from the colored portion).

are always satisfied. These are called *Maekawa's theorem* and *Kawasaki's theorem*.

Maekawa's theorem: The number of mountains and valleys differs by two.

Kawasaki's theorem: The alternating sum of incident angles is 0.

Figure P.3 presents an example of a crease pattern appearing on a flat-folded sheet when it is unfolded. It meets the above two theorems. Let us fold sheets in various ways, and then unfold them to observe that the two theorems are always right. The bird crease pattern in Figure P.2 also meets the theorems at all the intersections.

Note that these theorems are necessary but not a sufficient condition. A flat-foldable crease pattern always meets these two conditions, but meeting these conditions does not always make flat folding possible. There are crease patterns locally but not globally foldable because of a collision. To understand the situation of "partly but not globally foldable," let us look at the simple crease pattern in Figure P.4a. Trying to fold the two valleys flatwise fails because one folded section hits the other. On the contrary, as shown in Figure P.4b, by having the fold lines a bit dislocated from that as shown in Figure P.4a, the paper can be folded without collision. The crease pattern in Figure P.4c may look a little complicated, but it fully meets the above conditions and is foldable globally.

P.4 Tessellation and Twist Folding

Tessellation is one way of creating tiling patterns by folding paper as shown in Figure P.5.

Periodic patterns on a tessellation look like a mosaic. You may enjoy the shades against the light. This folding technique is seen in pleats in clothing and has had many known

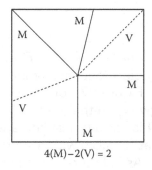
$$4(M) - 2(V) = 2$$

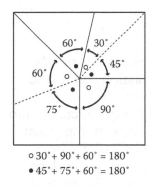
$$\circ \, 30° + 90° + 60° = 180°$$
$$\bullet \, 45° + 75° + 60° = 180°$$

Figure P.3   Explanation of Maekawa's theorem (left) and Kawasaki's theorem (right).

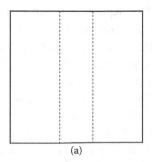
(a)

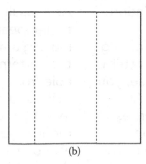
(b)

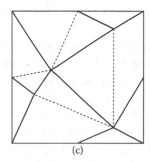
(c)

Figure P.4   (a) Flat-foldable crease pattern. (b and c) Flat-foldable crease patterns.

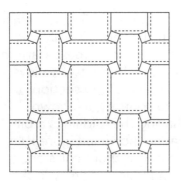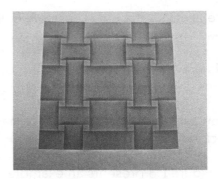

Figure P.5 Tessellation: from the left—crease pattern, photo of the folded paper, and translucent silhouette against the light.

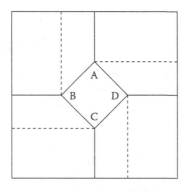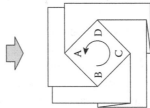

Figure P.6 Square-based twist folding.

patterns since long ago. Shuzo Fujimoto's extensive studies in tessellation conducted in the 1960s are still highly valued.

The concept of "connecting structures into a bigger shape" in tessellation is important for origami design. Chapter 3 of this book explains the making of an origami work having connected multiple three-dimensional shapes on one sheet of paper based on the concept of tessellation. Chapter 6 introduces different forms in an advanced tessellation structure.

Tessellation often uses the "twist folding" shown in Figure P.6. For the crease pattern (left) in Figure P.6 to be folded flatwise, you need to do it as you twist the sheet. At this time, the center square turns 90 degrees. Normally, a sheet is folded straight on and on. Twist folding requires you to move the fold lines all together while twisting at the center.

The twist fold in Figure P.6 can be connected to its mirror inversion side-by-side as in Figure P.7 (top). Then, inverting it vertically gives you the pattern in Figure P.7 (bottom). This has a big closed area in the center but is flat foldable with no problem.

As in Figure P.8, a twist-fold crease pattern can be parallelly shifted and connected to the original pattern. At this time, the fold line's mountain and valley signs must be inverted. As a result, the square area of the twist fold appears on the backside. By carefully looking at the crease pattern and connecting the patterns, you can extend the shape foldable from one sheet of paper.

Placing of square twist folds leads to different tessellation patterns. Besides squares, regular triangles and regular hexagons can also be tiled closely on a plane. So, each of the twist folds in Figure P.9 can also be placed and connected similarly as before.

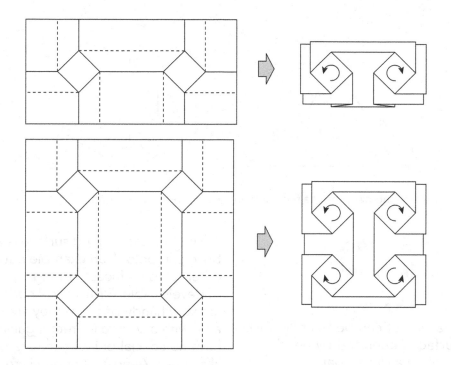

Figure P.7   Connection of twist folds.

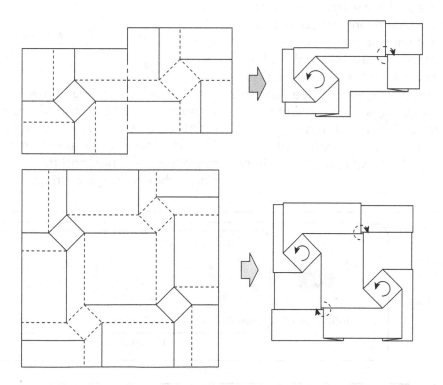

Figure P.8   Connection of twist folds with mountains and valleys inverted.

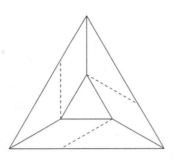
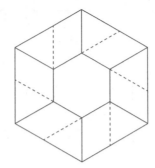

Figure P.9   Basic twist-folding patterns of a regular triangle and a regular hexagon.

P.5 Curved Surfaces Bendable from One Sheet

Paper is not elastic but can be bent flexibly. The curved surface generated by bending a sheet is called a *developable surface*. There are three developable surface types: *cylindrical surface*, *conical surface*, and *tangent surface*. These are subsets of a curved *ruled surface* represented as the locus of a straight line moved in space. A cylindrical surface consists of straight lines "in parallel," a conical surface "intersecting at one point," and a tangent surface "touching the space curve." It is a bit difficult to understand them only through reading. Figure P.10 depicts the classification of curved surfaces (only those representable by two variables).

The leftmost curved surface is a hyperbolic paraboloid, an example out of many other ruled surfaces. Among the three types of developable surfaces, cylindrical surfaces are very handy because they are easily foldable with a square (or rectangular) sheet and have no special point concentrated with lines like cones. Almost all origami creations with curved surfaces introduced in this book consist of sets of cylindrical surfaces.

P.6 Solids Made from One Sheet without Cutting

On the surface of a solid made from paper without cutting, the sum of angles around any point needs to be 360 degrees (Figure P.11). The developable surfaces introduced above meet this condition and thus are made from a single sheet without cutting.

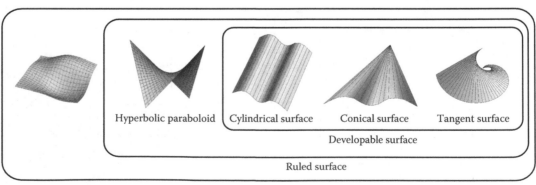

Figure P.10   Classification of curved surfaces.

A sphere or torus (donut-shaped solid) as in Figure P.12 requires cuts to be made on a single sheet. This is because the sum of angles around a certain point is less than 360 degrees (for a sphere) or greater than 360 degrees (for a torus).

Now, should we give up on making a curved surface as shown in Figure P.12 with one sheet of paper? The answer is "No." A piece of candy (sphere) is wrapped with one square sheet as photographed in Figure P.13. Once creased, even a sheet

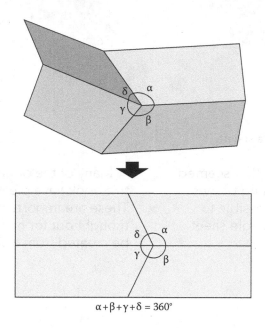

$$\alpha + \beta + \gamma + \delta = 360°$$

Figure P.11 Condition to eliminate the need for cutting paper.

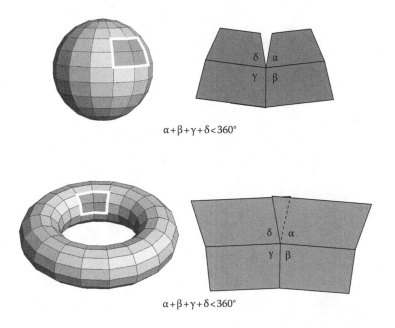

$$\alpha + \beta + \gamma + \delta < 360°$$

$$\alpha + \beta + \gamma + \delta < 360°$$

Figure P.12 Curved surfaces made with cut.

Figure P.13   Square paper wraps a sphere.

of paper can make a shape that seemed impossible. Precise computing of these crease positions makes it possible to fold a sphere even from a single sheet of paper.

Many of the origami creations introduced in this book have pleats and projections outside. These are important indispensable "creases" thought out for naturally impossible shapes to be created from a single sheet of paper.

Tips for Making Origami Creations Introduced in This Book

I would like to end this Prologue with some tips to help you make the origami creations introduced in this book.

GET THE CREASE PATTERN

This book has five "Let's Make It" projects providing photo-based explanations together with relatively simple crease patterns. You may want to use each crease pattern photocopied on a thick sheet or print out the data downloaded from the Internet. Visit my webpage at http://mitani.cs.tsukuba.ac.jp/book/3d_origami_art/.

In addition to these, all the crease patterns in this book are accessible. Try out any shape that intrigues you.

The recommended type of paper is a rather thick sheet capable of memorizing the folded shape. Many of the creations in this book are made using Tanto paper, which is easily available in Japan, and can be purchased on the Internet.

Each crease pattern in this book may be too small as is. I recommend enlarging them when photocopying or printing. For easier work, enlarge to A4 size for a simple crease pattern and A3 (11 × 17 inches) for more complex patterns.

Plain origami paper is colored on one side, so it is clear from the diagram as to which side is the front and which is the back. However, some origami creations in this book may be a bit confusing. Basically, in this book, mountain lines are folded as convex when viewed from outside. This rule is reversed for the "Let's Make It" crease patterns, meaning that mountain lines are folded as convex when viewed from the inside. This is because the finished work looks more beautiful when the printed side is inside.

On crease patterns in this book, the ● symbol means that the two sides having this symbol need to be glued together. Add an extra glue tab as necessary.

PRECREASE THE LINES

The origami creations in this book differ greatly from those in general origami books. With some, you need to fold along curved lines. No creation is ever easy. The more fold lines, the longer it takes to complete and the more difficult it is to achieve a beautiful finish. It took me years of experience to find my way of making these creations by using a "cutting plotter" (explained later) and folding up with time and care. It may not go smoothly in the beginning, but keep at it. Eventually, you will achieve a satisfactory outcome.

To finish your work beautifully, it is important to precrease the lines firmly in the beginning. Without this, it is impossible to fold a sheet at the intended positions. Sharp precreases make the workpiece neat at each step in the succeeding folding process. It can be said that precreasing counts the most in the process.

If you do not have a plotter, use a sharp pointed tool, such as a stylus, to press firmly along the fold lines. Avoid printing the fold lines directly on the production sheet, as they spoil the finished work by showing the printed lines. Instead, place the crease pattern–printed copier paper on the production sheet and then make precreases from over the copier paper.

A ruler is good enough to precrease straight lines but not curved lines. Still, avoid free-hand precreasing of curves. A curve ruler is a nice idea. The best way is to cut out the curved portion from a thick sheet and use it as the special ruler, because curves on a crease pattern are often in the same shape. The "Let's Make It" projects provide this curve ruler template of the crease pattern for your convenience.

For gentle curves, one method is using a straight ruler to approximate them with polygonal lines. The outcome is a bit nicer than drawing freehand.

The crease patterns for the creations in this book are available from my website http://mitani. cs.tsukuba.ac.jp/book/3d_origami_art/. If you have an available cutting plotter, load the data to the plotter software and let the machine make sharp precreases.

Cutting plotters are machines that automatically cut paper by controlling the cutter blade and paper feed. As the name suggests, a cutting plotter is intended for making cutouts; it can also be used to make precreases by lightly tracing the surface with the cutter blade at a controlled projection and force, or it may also be set up to finely perforate the fold lines. You may also prepare a precrease-dedicated blade by dulling a cutter blade. (Certain plotters offer a "marking" feature capable of precreasing, but they are very expensive.)

One example of an affordable small cutting machine is the Silhouette Cameo, which is capable of handling up to A4-size sheets (or letter size). Unfortunately, A4 size is not big enough to process many of the origami crease patterns in this book. A cutting plotter supporting up to A3 size is desirable, but it costs more. An alternative is dividing the crease pattern into A4-size portions and gluing them together later.

FOLDING PROCESS

Once you make sharp precreases on the sheet, you are more than halfway to a beautiful finish. Still, the succeeding folding process is not an easy task because it is difficult to fold curved lines until you get used to it. Folding on a desktop gives a straight fold. To fold a curved line, hold the sheet up with both hands and apply the intended fold as you warp the sheet. Fold the lines as sharply as possible so that the sheet memorizes the fold state.

Even a straight fold line is hard to accomplish if it is near the center, away from the edge. Avoid making unnecessary folding traces. Carefully fold only the necessary portions. Try twisting the entire sheet, which gives you an unexpectedly successful result.

Many of the origami creations in this book are hard to fold even though they are composed of straight lines. Some have a twist closing, some have a "twist" structure as in tessellations, and some have several solids connected. There are many cases in which you have to fold all the creases together at the same time. All we can do about this is practice and keep working at it. If you feel excessive force is being applied to the sheet, unfold it, firmly recrease the fold lines, and then try again. Force folding may unintentionally collapse a portion.

A crease pattern of the wrong size (too small or too large) cannot be comfortably handled. Use a crease pattern size that is controllable with both hands.

FINISHING

Sometimes gluing may be necessary to stabilize the final shape. It is difficult to fold paper or keep it exactly at the intended angle. Normally, the folded paper opens gradually if left as is. Most origami creations in this book have portions where the paper overlaps. Fixing such portions with wood glue stabilizes the whole shape. For a cylindrical creation, you need to join one edge to the other. In some cases, you need to allow for glue tabs in advance. Many of the cylindrical creations in this book have rectangular crease patterns. One solution to make the glued portion less visible is to reposition the edges so that the sheet can be glued exactly at a fold line.

TAKING PHOTOS

Take a photo of your finished work. Although it depends on your preference, a work made from white paper results in beautiful photographs in three-dimensional black-and-white shading. Some works may look great when lit from the back or when an LED lamp is placed inside. Firmly folded lines produce a sharp contrast, making for a good-looking work.

These are the tips for making a beautiful work. Again, the origami creations in this book are quite unusual and tougher to fold than you think. Never give up. Try many times, and then you will find yourself capable of doing it wonderfully. Now let us get down to folding and enjoying the magic of a sheet of paper metamorphosing into a beautiful piece.

Appendix: Simple, but Hard-to-Fold Crease Patterns

Below are two apparently simple crease patterns composed of only horizontal and vertical lines. Both look easily foldable because you only have to mountain fold the solid lines and valley fold the dashed lines.

Are they really easy to fold?

They do not look that different, but the left is easily foldable and the right very difficult to fold (found by Jason Wang).

After folds, a sheet overlaps in layers at some portions. The left crease pattern can be folded sequentially from the bottom into a multilayer-like structure. On the right crease pattern, however, the layers overlap cyclically like rock–scissors–paper, and thus you cannot determine which side should be the top.

This may be difficult to imagine only through reading. Try these crease patterns on an enlarged photocopy or drawn on graph paper.

Crease patterns that are simple in appearance but difficult to fold do exist. This book contains many origami creations with less fold lines but that are unexpectedly tricky.

There are many origami shapes that appear simple but are far from it.

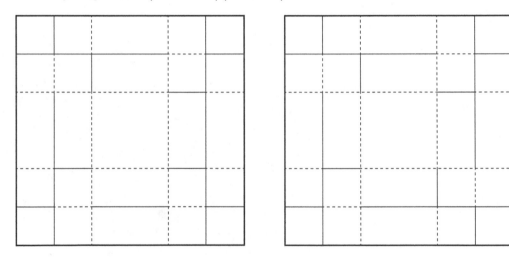

Author

Jun Mitani is a professor of information and systems in the Faculty of Engineering at the University of Tsukuba, Tsukuba, Japan. He is engaged in computer graphics–related research.

Born in 1975 in Shizuoka, Japan, he graduated with a doctor of engineering degree from the University of Tokyo, Tokyo, Japan, in 2004.

After working as a postdoctoral researcher at RIKEN in 2005 and as a lecturer in the Department of Computer Science at the University of Tsukuba in 2006, Mitani has been in his present post since 2015. He has served as councillor in the Japan Origami Academic Society. Involved in origami research as a PRESTO researcher at the Japan Science and Technology Agency from 2006 through 2009, Mitani is now continuously researching computer-aided origami design techniques. His early paper crafting and fascination with computers has led to the current research theme. His main books are *Spherical Origami*, 2009 and *3D Magic Origami*, 2010 (Futami Shobo Publishing).

Chapter 1

Axisymmetric 3D Origami

This chapter introduces you to the design techniques for axisymmetric three-dimensional (3D) origami shapes such as those in Figure 1.1. Some look as if wrapping a solid and others partly sticking to a solid, both with external projections. The following explains four types combined with two solid wrapping types and two external projection types.

1.1 Four Basic Types

The polyhedron on the left in Figure 1.2 is formed by rotating the polygonal line on the right (called the *section line*) around the vertical dash-dot axis at 60-degree intervals and connecting the vertices. In computational geometry (CG) and computer-aided design (CAD), this polyhedron is generated with a method called *rotational sweep* (the base is intentionally excluded in Figure 1.2). This chapter targets origami shapes having a structure that appears as if it wraps an axisymmetric polyhedron like in Figure 1.1 with one sheet.

There are various ways of wrapping a solid. We are looking at two methods in Figure 1.3: (a) overlaying ("cone type") and (b) cylindrical, surrounding from outside ("cylinder type"). Each type corresponds to the candy wrapping method in the Prologue, Figure 0.13. The solid does not have to be completely wrapped and can have its top and bottom left open. Use a regular polygonal sheet for the cone type and a rectangular sheet for the cylinder type.

The solid can have two types of external projections: a thin pleat called a flat pleat, and a thick solid pleat (triangular cross section) called a 3D pleat (Figure 1.4).

1.2 Basic Crease Patterns

As content common to the four different types of 3D origami, let's look at a rotationally swept solid and its crease pattern. In Figure 1.5, the solid on the left can be made by putting six of the parts on the right together.

For the cone type, the crease pattern components are placed on radially on a plane around the center as in Figure 1.6. The sheet used will be a regular polygon. Pleats are made from the margins outside the parts.

For the cylinder type, the crease pattern components are aligned at equal intervals as in Figure 1.7. The sheet used will be a rectangle. Pleats are made from the margins outside the parts.

Pleats are formed by adding new fold lines in the margins on the pattern. The way of adding fold lines varies depending on whether the pleat shape is flat or a triangular cross-sectional solid.

The following sections explain the four 3D origami types varying with the combination of pleat and wrapping types.

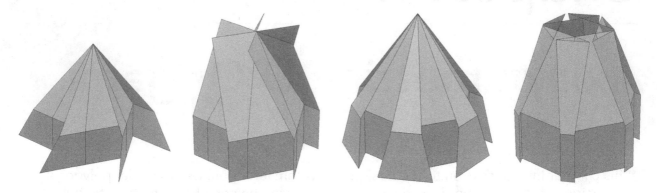

Figure 1.1   Axisymmetric 3D origami shapes presented in this chapter, from the left, flat-pleat cone type, flat-pleat cylinder type, 3D-pleat cone type, and 3D-pleat cylinder type.

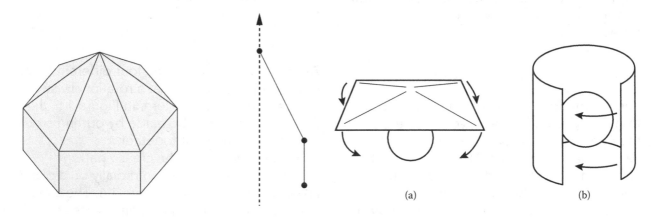

Figure 1.2   Rotationally swept axisymmetric solid (left) and polygonal line (right) used for making the solid.

Figure 1.3   Two ways of wrapping a solid: cone type (a) and cylinder type (b).

(a)

(b)

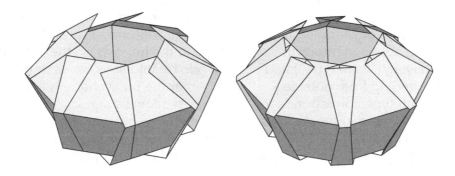

Figure 1.4   Two pleat types: flat pleat (left) and 3D pleat (right).

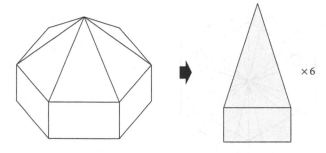

Figure 1.5 Rotationally swept solid and its components.

Figure 1.6 Components placed radially.

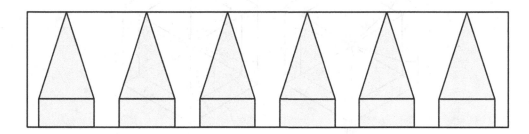

Figure 1.7 Components placed horizontally at equal intervals.

1.3 Flat-Pleat Cone Type

Shapes of this type have external flat pleats and look as if they are covering a solid from the top (Figure 1.8).

The crease pattern for this type can be created as follows. First, place the crease pattern components at equal intervals radially with their tips at the center as in Figure 1.9a. Second, add fold lines in the margins as in Figure 1.9b, so as to connect each vertex of the outer regular polygon to its center and to extend each part's horizontal line segment. Third, as in Figure 1.9c, erase the polygonal contour lines from each component (left-side polygonal lines are erased in the example). Last, assign mountains and valleys as in Figure 1.9d. A mountain fold line (solid)

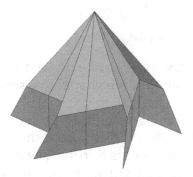

Figure 1.8 Flat-pleat cone–type solid shape.

makes a convex when viewed from outside and a valley fold line (dashed) a concave.

Through the above sequence, the flat-pleat cone–type crease pattern can be created for various axisymmetric solids.

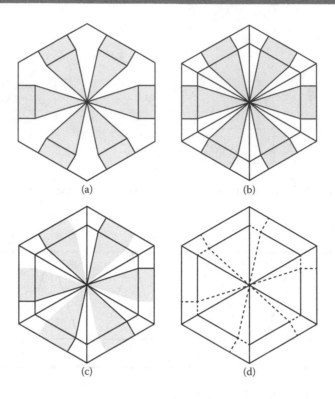

(a)　　　　　(b)

(c)　　　　　(d)

Figure 1.9　Flat-pleat cone–type crease pattern structure.

1.4 Flat-Pleat Cylinder Type

Shapes of this type have external flat pleats and look as if they are wrapping a solid from the side (Figure 1.10).

The crease pattern for this type can be created as follows. First, place the crease pattern components horizontally at equal intervals as in Figure 1.11a. Second, add fold lines in the margins as in Figure 1.11b, so as to add a vertical center line in each margin between the components and to extend each part's horizontal line segment. Third, as in Figure 1.11c, erase the polygonal contour lines from each component (left-side polygonal lines are erased in the example). Last, assign mountains and valleys as in Figure 1.11d. A mountain fold line (solid) makes a convex when viewed

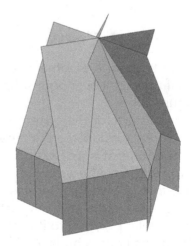

Figure 1.10　Flat-pleat cylinder–type solid shape.

from outside and a valley fold line (dashed) a concave.

Through the above sequence, the flat-pleat cylinder–type crease pattern can be created for various axisymmetric solids. Unlike the

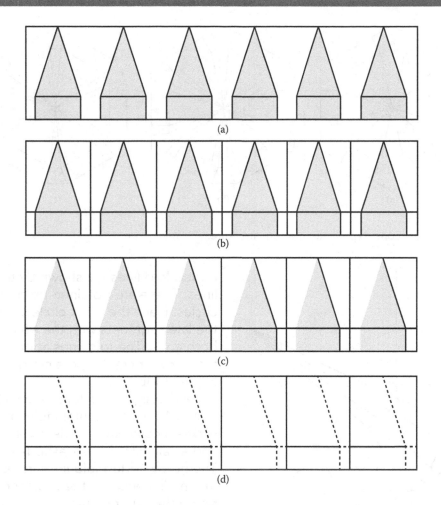

Figure 1.11   Flat-pleat cylinder–type crease pattern structure.

cone type, the pleat size can be adjusted by changing the margin width (i.e., the part placement interval). A larger interval makes a bigger pleat. A smaller interval makes a smaller pleat.

1.5 3D-Pleat Cone Type

Shapes of this type have external 3D pleats (projections) and look as if they are covering a solid from the top (Figure 1.12).

The crease pattern for this type can be created as follows. First, place the crease pattern components at equal intervals radially

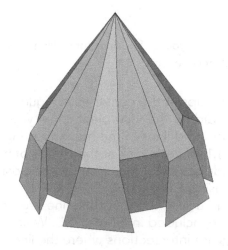

Figure 1.12   A 3D-pleat cone–type solid shape.

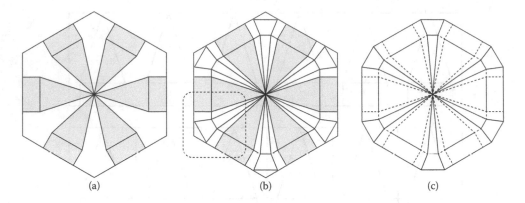

(a) (b) (c)

Figure 1.13   A 3D-pleat cone–type crease pattern structure.

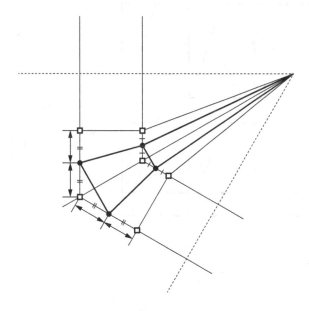

Figure 1.14   Layout of fold lines for making projections on 3D-pleat cone type.

with their tips at the center as in Figure 1.13a, like you did for the "flat-pleat cone type." Next, add fold lines in the margins as in Figure 1.13b. Let's look at the details of where these additional fold lines should be positioned using Figure 1.14.

In Figure 1.14, the thin solid line segments are those included in Figure 1.9b. White squares are intersections where the line segments in Figure 1.9b meet. The newly

added fold lines are shown by thick, solid lines. These are obtained by connecting black circles. Place the black circle so that it bisects the line segments with the white square on both ends. The fold lines added in this way are used for creating the crease pattern as in Figure 1.13b.

Last, assign mountains and valleys as in Figure 1.13c. A mountain fold line (solid) makes a convex when viewed from outside and a valley fold line (dashed) a concave. The newly added fold lines (radial lines) are mountains because the projections should be on the outer surface.

1.6 3D-Pleat Cylinder Type

Shapes of this type have external 3D pleats and look as if they are wrapping a solid from the side (Figure 1.15). When the solid has a pointed tip as we saw above, 3D pleats hit against each other as in Figure 1.15 (left). So the shapes that can be handled are limited. My explanation here targets the shape in Figure 1.15 (right) with its tip truncated.

The crease pattern for this type can be created as follows. First, place the crease pattern parts side-by-side as in Figure 1.16a. Next, add fold lines in the margins as in Figure 1.16b. Let's look at the details of where

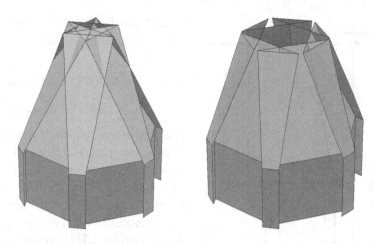

Figure 1.15 A 3D-pleat cylinder–type solid shape (projections hit against each other in the left).

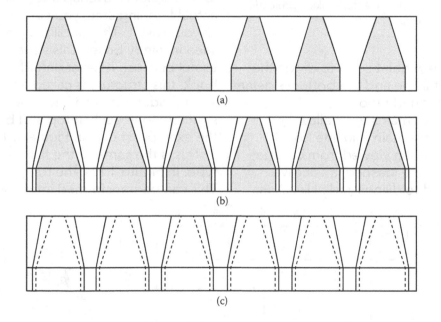

(a)

(b)

(c)

Figure 1.16 A 3D-pleat cylinder–type crease pattern structure.

these additional fold lines should be positioned using Figure 1.17.

In Figure 1.17, the thin solid line segments are those included in Figure 1.11b. White squares are intersections where the line segments in Figure 1.11b meet. The newly added fold lines are shown by thick solid lines. These are obtained by connecting black circles.

Place the black circle so that it divides the line segment with the white square on both ends at the ratio a:b = cosθ:1. Set θ to 180°/N, where N is the number of circumferential segments used for making the solid and equals the number of components placed. Therefore, N is 6 in this example. Though Figure 1.17 has the symbols a and b only at the upper side,

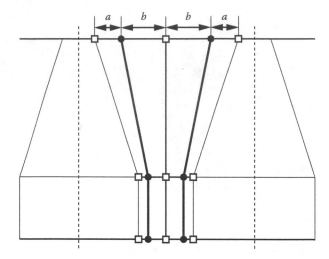

Figure 1.17 Layout of fold lines for making projections on 3D-pleat cylinder type.

also place the new point so as to divide the middle horizontal line and the bottom side at the above-mentioned ratio.

Last, assign mountains and valleys as in Figure 1.16c. A mountain fold line (solid) makes a convex when viewed from outside and a valley fold line (dashed) a concave. The newly added fold lines (radial lines) are

mountains because the projections should be on the outer surface.

1.7 "Twist Closing" for Closing a Solid

The flat-pleat cylinder type can make the shape of wrapping a solid from the side with the tip tightly closed as in Figure 1.18a. However, a paper shape in this state is unstable and needs gluing to prevent it from opening up.

The solution is the dark gray portion as in the Figure 1.18b crease pattern. Extend certain fold lines outside to make allowance for the sheet to overlap after folding. In the actual finishing step, you close the workpiece as you twist it. This stabilizes the shape. As seen in candy boxes, this structure is very useful because it never unfolds itself. In this book, this structure is called a *twist closing*. The cylinder type can have the twist-closing structure, one each at top and bottom as in the example shown in Figure 1.18b.

This is the same as the flat-pleat cone type. In Figure 1.19a, the top is tightly closed. This can be transformed into a twist-closing

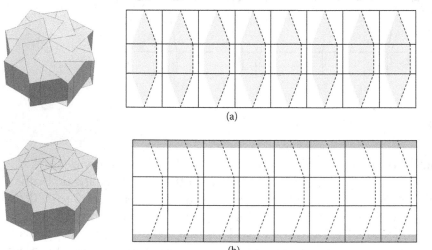

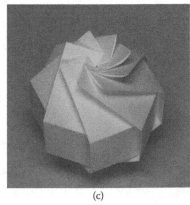

Figure 1.18 Twist-closing structure on flat-pleat cylinder type.

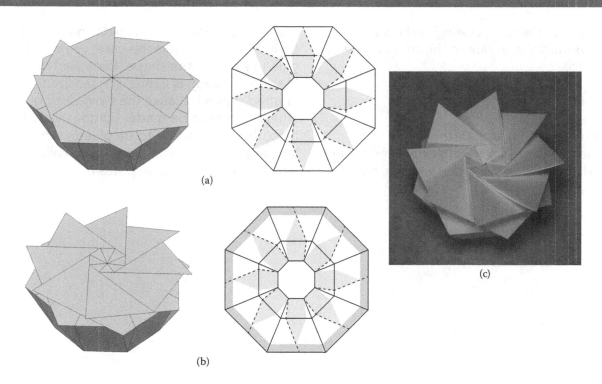

(a)

(b)

(c)

Figure 1.19   Twist-closing structure on flat-pleat cone type.

structure by extending fold lines outside to make a dark gray portion as in the
Figure 1.19b crease pattern.

However, the 3D-pleat cylinder and cone
types cannot form the twist closing because
the solid pleats hit against each other as in
Figure 1.15.

1.8 Solid with Curved Surfaces

A solid having smoothly curved surfaces
can be created by curving the cross section
of an axisymmetric solid. In Figure 1.20, for
instance, an arc on the left produces the flat-
pleat cylinder–type solid on the right. The
example here is circumferentially divided into
16 equal segments and is not an exact sphere
but has a "sphere-like shape." Each curved
surface patch has straight lines aligned equiv-
alent to the "cylindrical surface," one of the

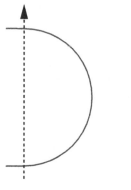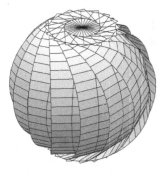

Figure 1.20   Three-dimensional origami with curved
surfaces.

developable surfaces. Making this shape with
a sheet is not so hard because the task is not
folding but bending to form a curved surface.

With an increased number of circumferentially divided segments, the solid becomes
more like a sphere, but it concentrates layers

of paper at the twist closing and results in an awkward shape due to thickness at the overlap.

1.9 Stabilizing a Shape

The twist-closing structure is useful to fix shapes because the sheet partly overlaps and stabilizes on the structure. However, some shapes are unstable after folds and may open up if left as is. If this is the case, use wood glue to stabilize the shape. Particularly for 3D pleats, gluing is a must to stabilize the finished shape because the twist-closing structure cannot be made. For the cylinder type, the sheet needs to be glued at the edge into a cylinder shape. So, you physically add a glue tab on the crease pattern.

Let's Make It (1) Eight-Flap Sphere

This 3D origami piece looks as if it is embracing a sphere, the simplest solid. To form a cylindrical shape, the sheet needs to be glued at the edge. The shape is stabilized by the twist closing at top and bottom. For curved lines, hold up the sheet and fold the curves as you roll the entire workpiece.

PREPARATION
- Prepare the crease pattern on the next page photocopied (a bit enlarged) on a thick sheet.
- Precrease all the fold lines using a hard, pointed tool. For curved lines, cut out the template and trace along its edge to precrease neatly.
- The printed side will be the inside of the solid.
- A "glue tab" is provided at the edge for closing cylindrically.

PATTERN SHEET

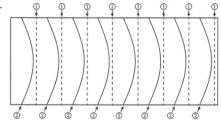

FINISHED PHOTO

PROCEDURE

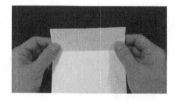

1. Fold all eight lines ① inward. Do not forget the line next to the glue tab.

2. Fold all eight curved lines ② to the opposite side of ① three-dimensionally as you roll the sheet.

3. Flip over the workpiece and re-crease firmly from the outside until it looks like, say, the back of a woodlouse.

4. Glue the paper edge and glue tab together to form a cylinder (the printed side is inside).

5. Wad up the workpiece so that the eight flaps overlap each other. If it does not work, re-crease the lines and try again.

6. Flip over and close the opposite side in the same way. Last, fix the shape of the flaps. Done.

"EIGHT-FLAP SPHERE" CREASE PATTERN

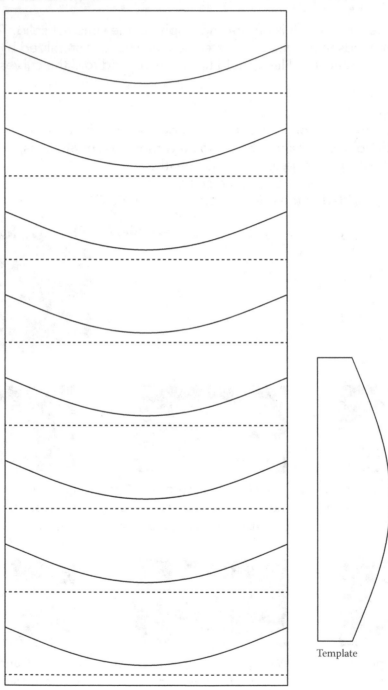

Template

The handy size is that the crease pattern's long side fits to the A4 sheet's longitudinal side. This crease pattern is downloadable from my webpage http://mitani.cs.tsukuba.ac.jp/ book/3d_origami_art/.

Let's Make It (2) Japanese Apricot Flower

This 3D origami is a flat-pleat cone type with a playful touch of contour on the crease pattern. Though the crease pattern looks like a flower, the finished work has a disk as its base. At the final step, you will see your work popping out like a flower. This simple, fun creation has fewer fold lines, and thus is not so hard to fold. Be careful and slow when closing the workpiece at the twist-closing structure.

PREPARATION

- Prepare the crease pattern on the next page photocopied (a bit enlarged) on a thick sheet.
- Precrease all the fold lines using a hard, pointed tool. For curved lines, cut out the template and trace along its edge to precrease neatly.
- The printed side will be the inside of the solid.

PATTERN SHEET

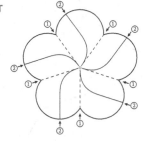

FINISHED PHOTO

PROCEDURE

1. Fold all five lines ① inward.

2. Fold all five curved lines ② outward, three-dimensionally as you roll the sheet.

3. Flip over and re-crease firmly from the outside.

4. Gently wad the five petals toward the center so that they overlap each other.

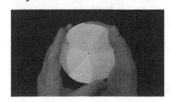

5. The workpiece becomes a circle when closed tightly.

6. Crease tightly from the outside to make it stable. Shape the whole form. Done.

"Japanese Apricot Flower" Crease Pattern

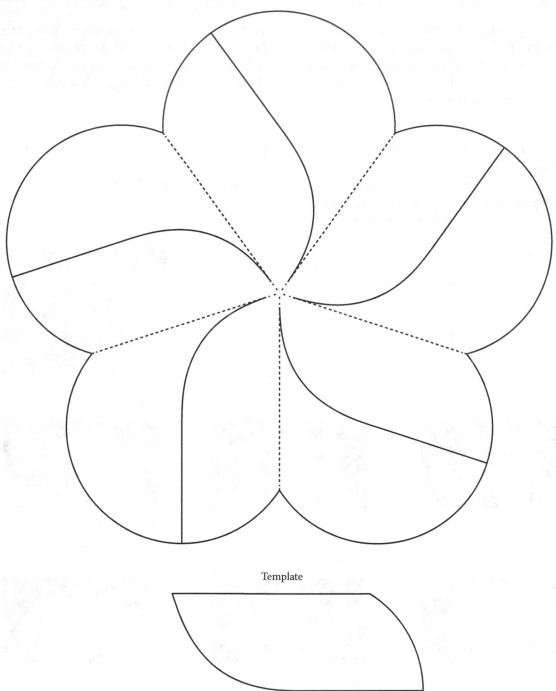

Template

The handy size is that the crease pattern's width fits to that of the A4 sheet in portrait. This crease pattern is downloadable from my webpage http://mitani.cs.tsukuba.ac.jp/book/3d_origami_art/.

Let's Make It (3) Whipped Cream

Consisting only of curved lines, this origami whipped cream will have a smoothly curved surface. Make good use of the template to crease the curved lines crisply. The folding process is also great fun as you see the workpiece graduating into a shape as you cup it with both hands to let the center lift up.

PREPARATION

- Prepare the crease pattern on the next page photocopied (a bit enlarged) on a thick sheet.
- Precrease all the fold lines using a hard, pointed tool. For curved lines, cut out the template and trace along its edge to precrease neatly.
- The printed side will be the inside of the solid.

PATTERN SHEET

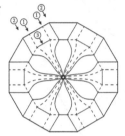

FINISHED PHOTO

PROCEDURE

1. Fold the curved line ① at six locations while folding the line ③ as well. The workpiece should look like the photo when viewed from the unprinted (outer) side.

2. Fold the curved line ② at six locations while folding the line ③ as well. The workpiece should look like the photo when viewed from the printed (inner) side.

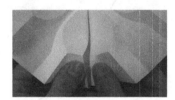

3. Gather the curved lines ② inward so that they get closer to each other. Crease tightly.

4. Shape the projections into whipped cream.

5. Once the workpiece is creased tightly to a degree, gently wad it up with your palms cupping it.

6. Tuck at the waist while pushing it against the desktop until the base gets flat. Glue to prevent opening. Done.

"Whipped Cream" Crease Pattern

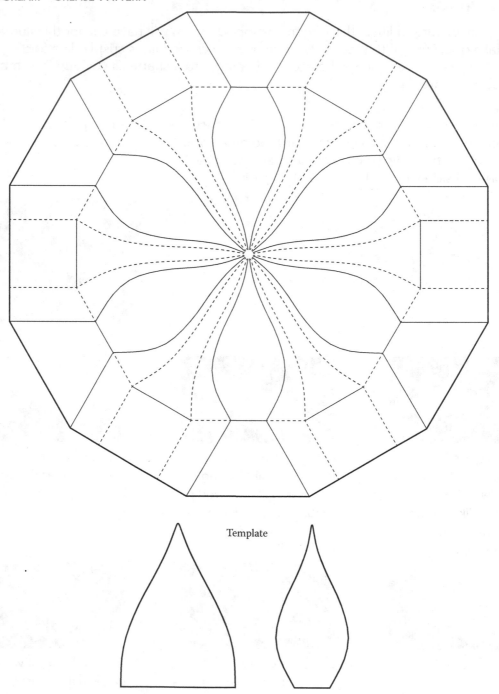

Template

The handy size is that the crease pattern's width fits to that of the A4 sheet in portrait. This crease pattern is downloadable from my webpage http://mitani.cs.tsukuba.ac.jp/book/3d_origami_art/.

16-Flap Sphere Origami

Compared with the eight-flap sphere in "Let's Make It" (see page 11, this 16-pleat sphere gives a finer impression. The crease pattern is very long because the surface area increases with more pleats.

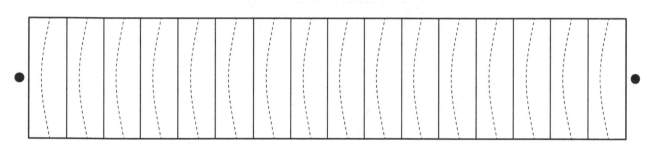

Egg Wrapping

This cute, flat-pleat cone–type creation is achieved by having an egg-shaped solid inside. The creation is stable without glue due to the twist-closing structure at the bottom.

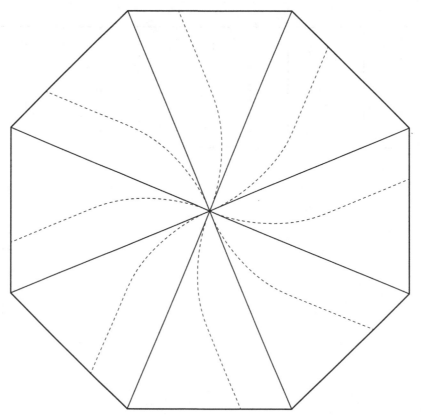

Pear

The structure is almost the same as the sphere on page 17. By changing the curved line that makes up the cross section, the creation is shaped into a cute pear. The extra, pasted stem may be a violation of the spirit of origami but is a charming point for this creation.

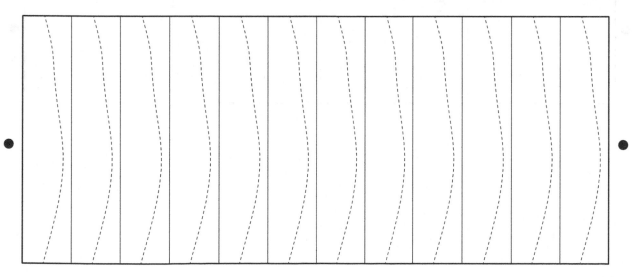

Water Wheel

In contrast to the idea of "wrapping a solid," this cylinder opens outward at both ends. The creation has a beautiful curved surface inside and a water wheel–like shape outside.

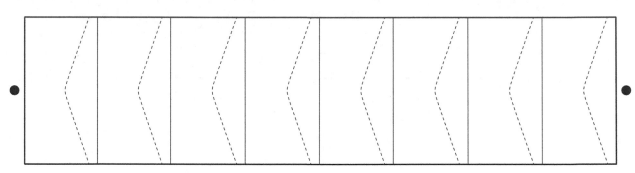

Gear Wheel

Using 3D pleats, this creation is shaped like a machine part. For the solid section line, the first and last segments are made horizontal. The result is the "flattened" ends of each 3D pleat. When folding, the pleats slightly hit against each other at the edge but can overlap successfully because the portion is folded flatwise.

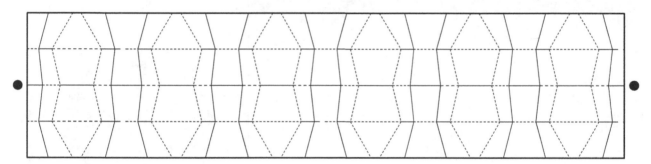

Appendix: 3D Origami Design Software

Though the 3D origami shape design method is clear, it is not an easy thing to actually draw a crease pattern. A calculator, rulers, and a pair of compasses may do the task, but why not use a computer? With commercially available CG or CAD software, you can make a 3D shape, arrange configuration faces onto a plane, and apply the methods in this book to draw a 3D origami crease pattern.

Still, trial and error in design is a tough thing as each step takes time. So, I have written several 3D origami design-dedicated software programs myself.

For instance, ORI-REVO has been developed for the easy design of the axisymmetric 3D origami shapes presented in Chapters 1 and 2. The software screen is structured as shown. You draw a polygonal line to make the cross section, and then the software automatically renders the finished image and the crease pattern. ORI-REVO supports the four origami types presented in Chapter 1 and is capable of changing parameters (e.g., pleat size, number of circumferential segments) and saving the crease pattern to a file. At this writing, ORI-REVO is downloadable free from the following URL. Give it a try. http://mitani.cs.tsukuba.ac.jp/ori_revo/

(The software is a Java application, and proper operation depends on your computer environment.)

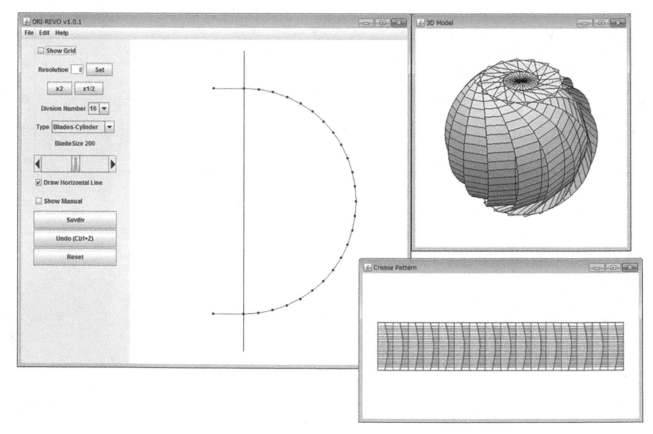

Chapter 2

Extension of Axisymmetric 3D Origami

Chapter 1 introduced you to origami shapes as if each were wrapping an axisymmetric solid inside. The principle is simple, and drawing a crease pattern is not very difficult. This simplicity leads to various applications, enabling you to create as many interesting shapes as you like. In this chapter, the flat-pleat type is evolved by various methods. Getting a bit away from the concept of "wrapping a solid" gives more freedom for creating a variety of shapes based on the same principle.

2.1 Connecting Two 3D Origami Shapes (Cylinder Type)

Figure 2.1 is the mirror image of the upper crease pattern in Figure 1.18. Its folded shape is an inverted mirror image. When one crease pattern is placed above the other, their vertical mountain fold lines are aligned at the same intervals, and boundary valley fold lines coincide with each other. For this reason, the fold lines are connected seamlessly throughout the two crease patterns and can actually be folded. The Figure 2.2 crease pattern is added with a twist-closing structure at the top and bottom ends. After folds, the shape becomes two connected solids as in the photo. The connected section is the twist-closing structure.

2.2 Connecting Different 3D Origami Shapes (Cylinder Type)

Let's take a close look at Figure 2.2. The upper and lower shapes do not have to be identical as far as the equally spaced vertical mountain fold lines and the slant valley fold lines for a twist-closing structure run seamlessly across the crease patterns. For the cylinder type, the pleat size can be set freely. Therefore, if the pleat size is adjusted so that the vertical mountains are equally spaced, as in Figure 2.3, then two different shapes can be connected up and down, or even a greater number of different shapes like the creation on page 30.

Normally, in 3D design where a polygonal line is rotated around the axis, the section line should not intersect at the axis. However, the solid in Figure 2.3 can be obtained by rotating the polygonal line in Figure 2.4 crossing the axis. Getting a bit away from the idea of "wrapping a solid," the section line can be determined freely without minding the intersection with the axis. Take a close look at Figure 2.4 and you can see that the section line crosses the axis three times. This means that the twist-closing structure appears at three sections: top, bottom, and center.

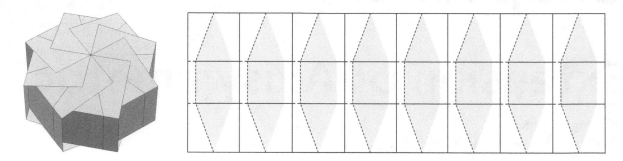

Figure 2.1   Mirror image of Figure 1.18 (a) crease pattern.

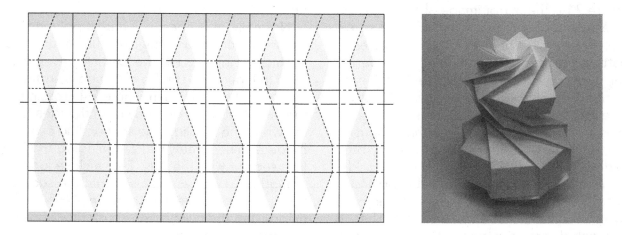

Figure 2.2   Connecting a shape and its mirror image.

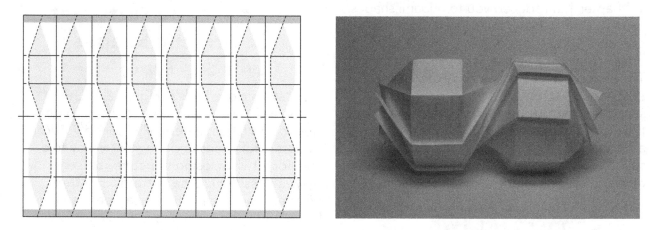

Figure 2.3   Two different solids connected (flat-pleat cylinder type).

The reason for the section line being able to cross the axis is the twist-closing structure, which allows the sheet to avoid hitting at the folded edges.

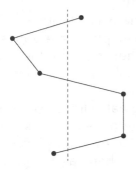

Figure 2.4   Section line for making the 3D origami in Figure 2.3.

2.3 Connecting Different 3D Origami Shapes (Cone Type)

Similar to the cylinder type above, the flat-pleat cone type also can make a 3D origami having multiple shapes connected by making use of the twist-closing structure. The section line in Figure 2.5a forms a flat-pleat cone–type 3D origami like the one in Figure 2.5b. The photo in Figure 2.5d is the real thing after folds. In Figure 2.5a, the section line crosses the axis twice, meaning that the twist-closing structure appears at two sections.

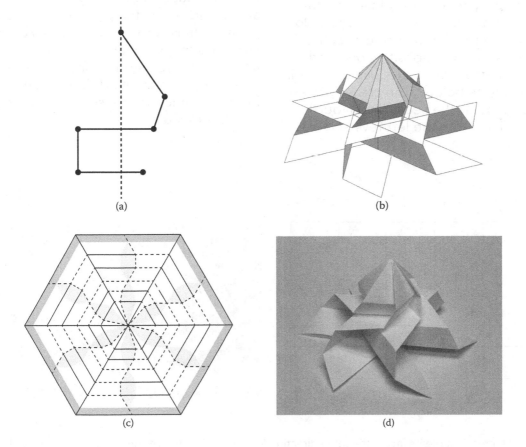

(a)

(b)

(c)

(d)

Figure 2.5   Two different solids connected (flat-pleat cone type). (a) Section line, (b) 3D model, (c) Crease pattern, (d) Photo.

In the crease pattern in Figure 2.5c, the light gray areas are the original solid components, telling you that the pleats account for most of this crease pattern. For the cone type, the solid components are radially placed and the margin is allocated for pleats. So, the more outward from the center, the larger the margin is on the crease pattern and the pleat on the folded solid.

2.4 Changing Pleat Orientation (Flat-Pleat Type)

The flat-pleat type has two pleat orientations: clockwise and counterclockwise. Not all pleats have to face the same direction. Some pleats can be oppositely oriented. The crease pattern in Figure 2.6a is made in the same way as before. The crease pattern in Figure 2.6b has the second and fourth valley fold lines from the left reversed with respect to the vertical mountain fold line. This simple operation can reverse the orientation of a pleat. As shown in the photo in Figure 2.6, a small alteration greatly changes the impression.

Note that the twist-closing structure cannot be made if some pleats are oppositely oriented from others because the pleats in this structure must always be overlapped in the same direction.

For the original solid, the component arrangement interval can be set freely. The components can even be placed with no space between them, as in Figure 2.7a. This zeros the pleat width at the portion where one component touches another. In this way, the orientation of pleats can be reversed above and below the portion where the pleat width is zero, as in Figure 2.7b. By using this in combination with the above method of reversing the orientation of some pleats, you can create different variations from the same original solid depending on how it is pleated.

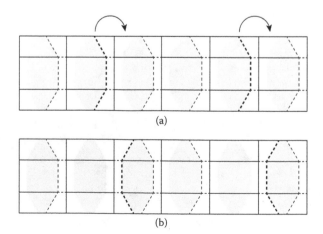

Figure 2.6 Shapes with pleats facing the same direction (crease pattern [a], left photo) and with some pleats facing opposite (crease pattern [b], right photo).

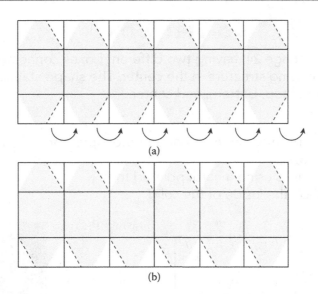

(a)

(b)

Figure 2.7 Crease pattern with pleats facing the same direction (a). Reversed orientation of pleats above and below the portion where the pleat width is zero (b).

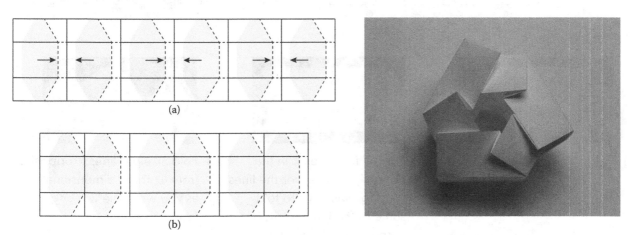

(a)

(b)

Figure 2.8 Crease pattern with same sized pleats (a). Example of a shape having some pleats of different sizes (crease pattern [b], photo).

2.5 Resizing Pleats (Cylinder Type)

We have seen that for the cylinder type, the pleat size can be changed by adjusting the component arrangement interval. Components do not have to be placed at equal spacing. For instance, enlarging or reducing the space between certain components changes the size of the pleat only at that section. Figure 2.8a is the same as Figure 2.6a. By zeroing the width of three pleats at every other portion, the crease pattern looks as it does in Figure 2.8b. The folded shape can be controlled by the pleat size. However, this can be done only for the cylinder type.

Let's Make It (4) Two-Tier Box

This shape is presented on page 24, having two different boxes connected. Folding may be a bit difficult for its twist-closing structure in the center. The shape stabilizes without gluing thanks to the twist closing at top, bottom, and center.

PREPARATION

- Prepare the crease pattern on the next page photocopied on a thick sheet.
- A3 size paper is desirable.
- Precrease all the fold lines using a hard, pointed tool.
- The printed side will be the inside of the solid.

PATTERN SHEET

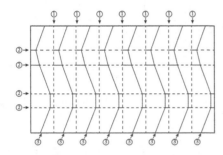

FINISHED PHOTO

PROCEDURE

1. Fold all eight lines ① inward.

2. Fold all four lines ② in the same direction as for the lines in ①. You will have to fold some of them reversely, but do not mind at this step.

3. Fold all eight lines ③ opposite from ①, three dimensionally as you wave the working faces.

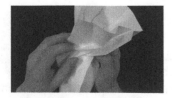

4. Fold down the central twist-closing structure as you twist the whole workpiece until the faces overlap each other.

5. When viewed from the side, the workpiece is twisted at the center.

6. Finish the two boxes one by one. Done.

"Two-Tier Box" Crease Pattern

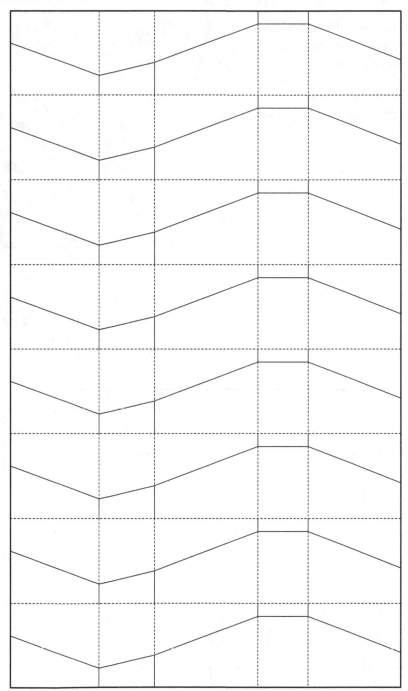

The handy size is that the crease pattern's long side fits to the A3 sheet's longitudinal side. This crease pattern is downloadable from my webpage http://mitani.cs.tsukuba.ac.jp/book/3d_origami_art/.

Three-Tier Octagonal Box

Three octagonal boxes in different sizes are connected vertically. The width of the crease pattern is adjusted by the pleat size for each different-sized box. This creation has four twist-closing structures—one each at top and bottom and two at the middle.

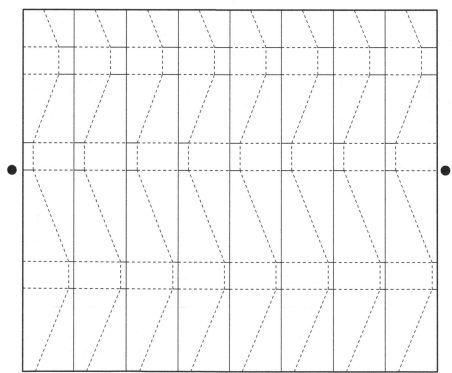

Twisted Tower

This flat-pleat–type origami piece is made from a solid having a stepwise cross section. At an exhibition, these creations were vertically stacked with a rod for stabilization. For this purpose, the crease pattern has a hole for the rod to go through. With this hole arrangement, the rod exactly runs through the central twist closing.

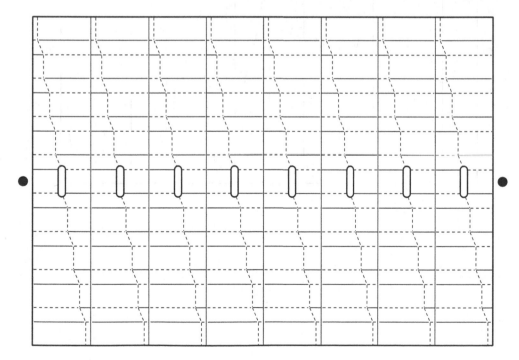

Two Connected Spheres

Two spherical bodies are connected. Prepare two identical sphere crease patterns, reverse one of them, and then place them up and down to be connected. The principle is very simple and so also is the crease pattern. Still, it is hard work to actually fold this creation due to its curved surfaces plus a twist-closing structure at the center.

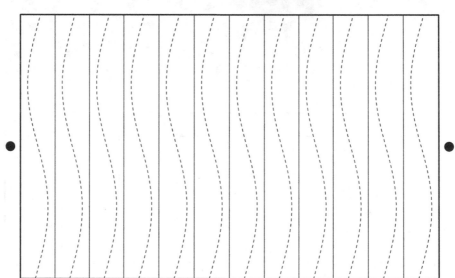

Bellows Cylinder

As explained on page 26, the pleat orientation is changed alternately in the circumferential direction. In the vertical direction, the pleat orientation is changed at the point where the pleat width is zero. The resulting creation has a tiling-like geometric pattern on the surface.

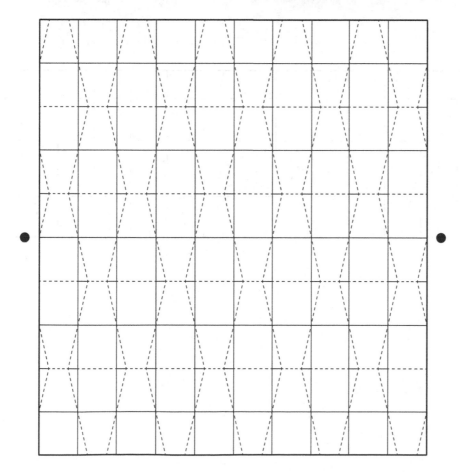

Sixteen-Pleat Spiral

This creation is a stack of four twist-closing structures alone. It has 16 pleats, and thus the crease pattern is filled with zigzag lines. The folded creation feels hefty, as most of the sheet is used for densely overlapping pleats.

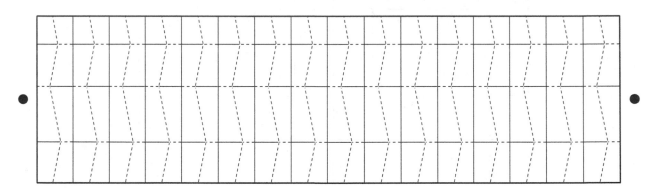

Sea Snails

When viewed from the front, these three have a regular decagon structure wrapping an identical solid. The pleat size is varied by changing the component arrangement interval on the crease pattern. The varying pleats give a great difference in impression of the finished shape. Pleats varying in length do not affect the central twist-closing structure.

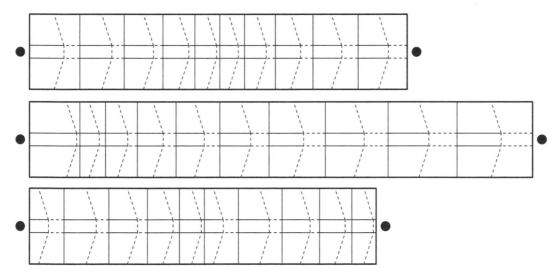

Chapter 3

Connecting Axisymmetric 3D Origami Shapes

For the cone-type axisymmetric 3D origami, the crease pattern inscribes in a regular polygon and the folded shape also has its bottom inscribed in a regular polygon. So, by connecting multiple crease patterns, you can make a state of multiple connected solids from one sheet. The cylinder type can be connected vertically, as explained in Section 2.1. The cone type can be connected horizontally to tile on a plane, regardless of whether the pleat shape is flat or 3D.

3.1 Connecting and Tiling 3D-Pleat Type on a Plane

On the flat-pleat type, pleats have "orientations." The 3D-pleat type is easier to handle because it is symmetric both on the folded shape and the crease pattern. So, I would like to start with the 3D-pleat type to explain how to connect multiple 3D origami shapes in the horizontal direction.

The regular polygons that can congruently tile a plane are triangles, squares, and hexagons only (Figure 3.1). The degrees of grid point (number of sides connecting to one intersection) are 6, 4, and 3, respectively. Putting polygons close on a plane is called "tiling." Here, we are going to arrange

3D-pleat cone–type 3D origami shapes on vertices of the regular polygonal tiling pattern.

For 3D origami shapes to be arranged on a plane, the last line segment of the cross section needs to be horizontal, as marked with a dotted oval in Figure 3.2a. This produces the horizontal, branch-like portions extending radially from the center, as marked with a dotted oval in Figure 3.2b. Hereinafter, the branch-like portion is called a *connecter*, and the 3D origami shape is called a *unit*.

Now, put a six-connector unit from Figure 3.2b onto a vertex of the regular triangle tiling pattern. The connectors exactly lap over the six sides connecting to the vertex. In this way, the identically shaped 3D origami units can be placed one after another on the regular triangle tiling pattern by connecting them at the connector, as schematically shown in Figure 3.2c. On the crease pattern, the connector is the section marked with a dashed oval in Figure 3.2d. Figure 3.2e is the crease pattern for the three connected units. The gray areas are the connectors connecting the units. Only three units are connected in this example, but as many as you wish can be connected in a similar manner. The connector can be in any length, allowing you to adjust the unit arrangement interval.

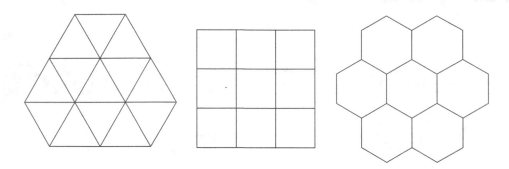

Figure 3.1 Regular triangle, square, and hexagon tilings.

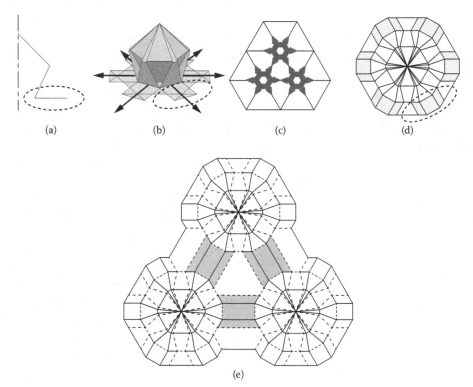

(a) (b) (c) (d)

(e)

Figure 3.2 The 3D-pleat–type connectors and how they are connected.

For a 3D-pleat–type unit, the six-, four-, and three-connector 3D origami can be placed on the vertex of the regular triangle, square, and hexagon tiling patterns, respectively, as in Figure 3.3. Figure 3.2 is an example for six connectors and Figure 3.4 for four connectors. A unit like this can be connected at the intersection of an orthogonal grid.

3.2 Connecting Flat-Pleat Type

On the flat-pleat type, pleats have orientations. So, to connect two units side by side, the pleats need to face oppositely on the neighboring units (Figure 3.5). This is the same as connecting twist folds on page XIII. (The neighboring unit can

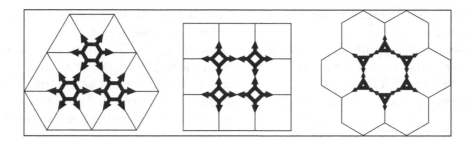

Figure 3.3　A 3D-pleat–type arrangement on tiling pattern.

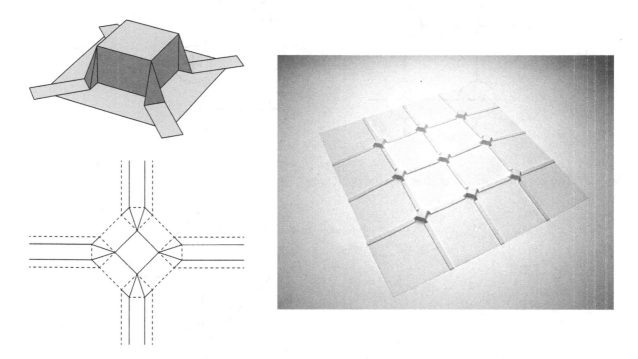

Figure 3.4　Four-connector units connected.

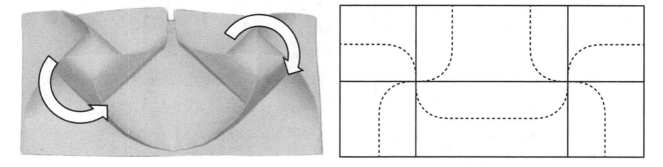

Figure 3.5　Connecting flat-pleat type.

face the backside by shifting the crease pattern as explained on page XIII.)

Therefore, the only tiling patterns that allow side-by-side arrangement are squares and hexagons, as in Figure 3.6. (The positive and negative signs represent clockwise and counterclockwise orientations, respectively.)

The number of connectors is four for squares and three for hexagons.

Similar to the 3D-pleat type, the flat-pleat–type 3D origami can also be arranged on a plane if the last segment of the cross section is horizontal. The cross section in Figure 3.7a makes the connector marked

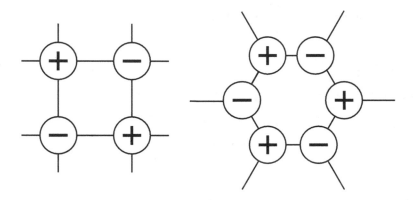

Figure 3.6   Two patterns capable of arranging flat-pleat type.

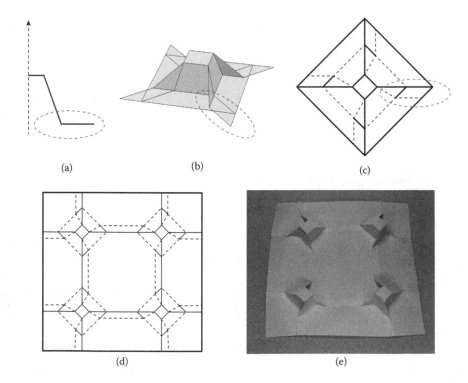

(a)   (b)   (c)

(d)   (e)

Figure 3.7   Flat-pleat type connectors and how they are connected.

with a dotted oval in Figure 3.7b. This connector extends in four directions on the crease pattern, as in Figure 3.7c, allowing the units to be placed and connected at the vertices of the square tiling pattern. Pleats have orientations. On the neighboring units, the pleat faces oppositely from the other.

3.3 Connecting Different 3D Origami Shapes

Identically shaped units can be connected by the method described earlier. Even differently shaped units can be connected as long as their connectors are the same width. For instance, the two 3D origami shapes in Figure 3.8 cannot be connected as is because their connectors are different widths. In this case, the connector width of either crease pattern can be made equal to the other by enlarging or reducing the entire crease pattern. The connector can be lengthened freely regardless of the crease pattern size.

By adjusting the connector width like this, the differently shaped solids can also be connected on a grid with no problem.

3.4 Making Use of Duality

We have seen tiling patterns consisting of regular polygons in identical shapes. As in Figure 3.9a, the arrangement for connecting units can also be made from the "dual" of a tiling pattern with different regular polygons (triangles and squares in this example).

For a tiling pattern made of regular polygons, the dual is obtained by connecting the centers of the neighboring regular polygons with a line segment. For instance, the dual of Figure 3.9a is Figure 3.9b. In the dual, the original polygons are swapped with intersections, and intersections are swapped with polygons.

Take a close look at the pattern of Figure 3.9b. Every vertex has edges connecting at equal angle intervals. Four edges gather in each area that was a square and three sides in each area that was a regular triangle. Therefore, by placing units with connectors

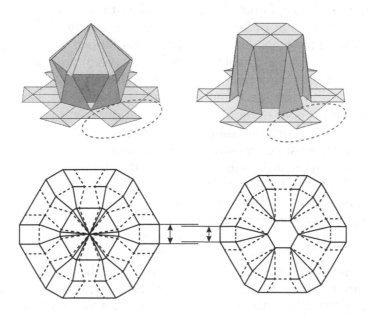

Figure 3.8   Two different 3D origami shapes.

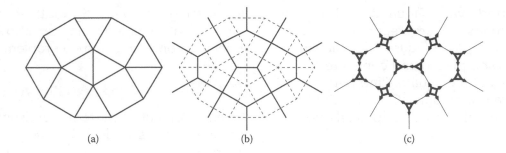

(a)　　　　　　　(b)　　　　　　　(c)

Figure 3.9　Regular polygon tiling and its dual.

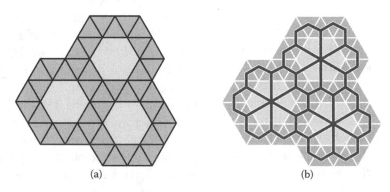

(a)　　　　　　　(b)

Figure 3.10　Tiling combined with regular triangles and hexagons, together with its dual.

equal to the number of sides onto the vertices, such units can be connected, as in Figure 3.9c.

Then, how is the pattern that has different regular polygons placed together with no space in Figure 3.9a made? A tiling, which is combined with multiple types of regular polygons under the restriction that "each vertex has the same type and number of regular polygons connected," is called Archimedean tiling (semiregular tiling). There are only eight known Archimedean tiling types, including those shown in Figures 3.9a and 3.10a, though we can create numerous patterns with freely tiled regular polygons. Using a known pattern is an easy way to start.

In the pattern in Figure 3.9a, a vertex has three regular triangles and two squares around it. In the pattern in Figure 3.10a, a vertex has four regular triangles and one hexagon around it. These are described

as (3,3,4,3,4) and (3,3,3,3,6), respectively, in Schläfli notation.

The dual entity of the Figure 3.10a tiling pattern is Figure 3.10b. The vertex, which is present in each area that once was a regular triangle, has three edges at equal angle intervals. The vertex, which is present in each area that once was a regular hexagon, has six edges at equal angle intervals. So, the three- and six-connector units can be connected to each other by being placed on each vertex.

3.5 Layering Dual Patterns

Figure 3.11a shows the crease pattern for two perpendicular connectors of different widths. This crease pattern is not flat-foldable as is, but can be folded if certain mountains and valleys are changed, as in Figure 3.11b. In other words, even an

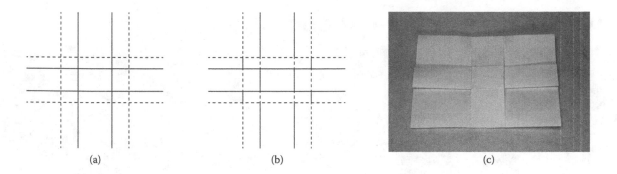

Figure 3.11 Perpendicular connectors.

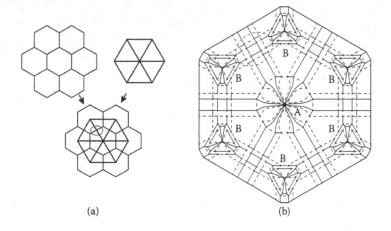

Figure 3.12 Layering of dual patterns (regular triangle tiling and regular hexagon tiling).

arrangement where the connectors cross each other can be made with one sheet if mountains and valleys are properly assigned. However, this is possible only when the connectors are perpendicular, but not for obliquely crossing connectors.

In Figure 3.12a, the regular triangle and hexagon tilings are in duality, as each vertex on one tiling laps over the center of each polygon on the other. In a regular polygon tiling, the edges included in the dual are perpendicular to those of the original regular polygon. Therefore, a regular triangle tiling–based 3D origami can be synthesized with a regular hexagon tiling–based one by simply layering their crease patterns, even though they are designed independently.

Figure 3.12b shows an example. In this crease pattern, six-connector Solid A is centered and surrounded by three-connector Solids B in connection with Solid A. Solid A is based on regular triangle tiling and Solid B on regular hexagon. Therefore, the connectors of Solids A and B cross perpendicularly to each other. By assigning mountains and valleys as shown in Figure 3.11b at the section where connectors intersect at right angles, the whole body can be created with one sheet (see page 46 for the creation done from the crease pattern in Figure 3.12a).

A square tiling is self-dual. In addition to the layering arrangement in Figure 3.12, the square tiling–based arrangement can be layered over each other.

Quadrangular Pyramid and Cube Assortment

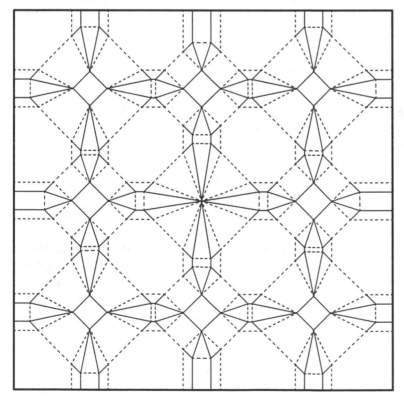

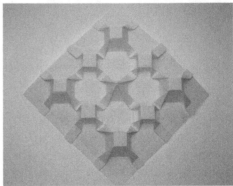

This creation has three different flat-pleat cone types of 3D origami shapes, nine in total. Each shape has four connectors. By making the connector widths identical, different 3D origami shapes can be connected like this.

Connected Whipped Creams

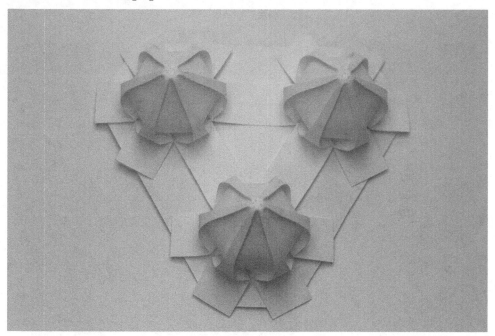

The whipped cream shape presented on page 15 has six connectors and thus can be connected to other whipped cream shapes by being placed at the vertex of a regular triangle tiling. This creation has three whipped creams connected together. Their fold lines are connected to each other, so you have to fold all three whipped creams at one time rather than sequentially.

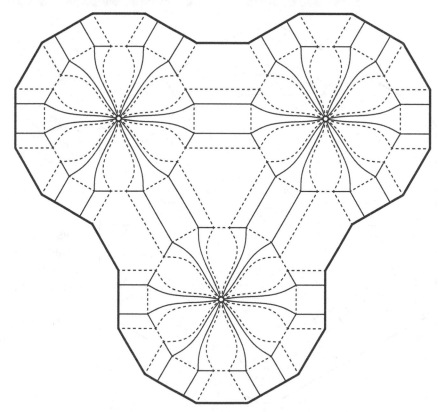

Merry-Go-Round of Whipped Cream and Trigonal Pyramids

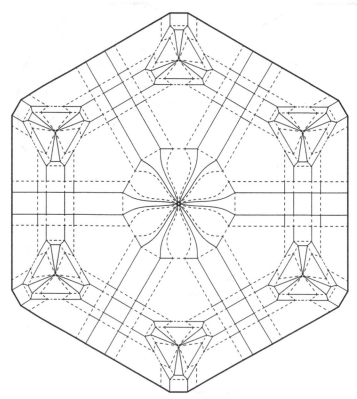

This creation is presented on page 43. It has only one hexagon element and has a layered structure of a hexagon tiling and a triangle tiling. The connectors can cross perpendicularly without affecting each other, allowing you to first make the whipped cream in the center.

Vortex

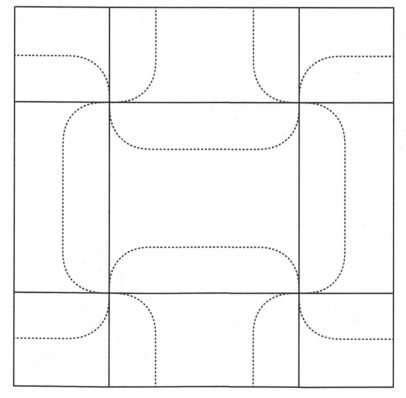

Four flat-pleat types with twist closing are connected. The crease pattern is very simple with a low number of fold lines. Still, it is hard work to actually fold this creation, because each shape has a twist-closing structure. This creation can be said to be the solid version of the tessellation presented on page XIII.

Triangular Plate Arrangement

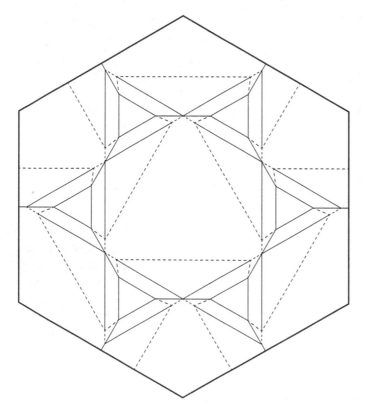

Regular triangle–based 3D origami shapes are connected together at the vertices of a regular hexagon tiling. Each has a twist-closing structure, thus making it a tough thing. At the final folding step, the six regular triangles come together toward the center as they turn around.

Extension of Connected Whipped Creams

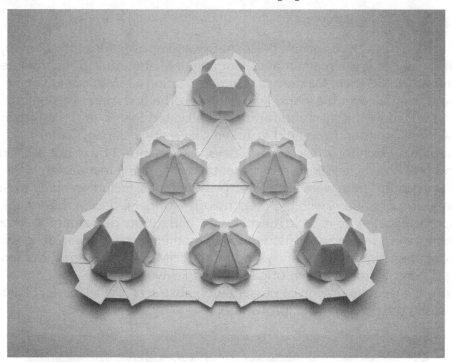

For the three whipped cream shape on page 45, three more units in a different shape are added externally. Theoretically, an unlimited number of shapes can be connected. However, connecting any more shapes will present many difficulties as it is very hard to fold lines away from the edge.

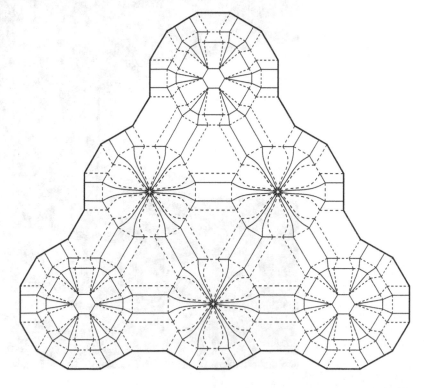

Appendix: OSME, The International Origami Conference

As an academic area of study, origami's basic theory and applications attract many researchers from a wide range of fields like mathematics, engineering, and education. The International Meeting on Origami in Science, Mathematics and Education (OSME) is an event at which such researchers meet from all over the world every 4 years. The first round was held in 1989 in Italy, the second in Japan, the third and fourth in the United States, and the fifth in Singapore. In August 2014, the sixth round was held at Tokyo University in Japan.

The sixth meeting, which took place over 3 days, was attended by about 300 origami researchers and enthusiasts, and featured presentations on the latest research results as well as enthusiastic discussions. There were about 140 presentations, including many topics that indicate the active fusion of origami with high technology—origami design theory, folding movement simulation, application to robotic mechanisms, and robots performing folding motions.

The participants were from various fields such as mathematics, mechanical engineering, biology, art, education, computer science, and more. OSME is a wonderful exchange event for people having different expertise with origami as the keyword.

At this writing, the documentation of presentation abstracts is available in PDF format from the 6OSME website (http://origami.gr.jp/6osme/).

Chapter 4

Making Use of Mirror Inversion

Mirror inversion is a geometric operation closely related to physical paper folding. The resulting shape after a mirror inversion that is applied to a section of a one-sheet shape will also be foldable with one sheet. Mirror inversion adds new folds to a shape.

4.1 Cone-Based 3D Origami

A cone can be rolled up from a fan-shaped sheet. (The cone's base is not considered here.) Pass a plane through the cone and flip the cone's upper part by the plane. The resulting shape, the upper part is mirrored, can also be made with one fan-shaped sheet, meaning that new fold lines are added on the crease pattern. See Figure 4.1 as an example. If you flip the cone's upper part around the horizontal plane across the cone, you get a shape like the one in Figure 4.1b. This shape has the original cone's upper part pushed in downward and thus can be made just by folding a sheet, without cuts. Now, flip only the section containing the cone's upper part again. You get the shape in Figure 4.1d where the upper part is looking up. This shape needs no paper cuts because the upper part is just folded back similarly as before. In other words, this is the same thing as adding new folds, and the shape can still be made from one paper. So, the fan shape that is necessary

for making the first cone is unchanged even after several mirror inversions.

The plane for mirror inversion can be parallel or oblique to the cone's base. Figure 4.2 is a shape resulting from repeated mirror inversions around oblique planes. The cross sections generated by foldback are slant. The cone vertex angle is set to 60 degrees, and thus the shape of the sheet is a semicircle.

4.2 Mirror Inversion on a Developable Surface

The mirror inversion technique for making new shapes applies not only to cones but also to all kinds of 3D origami.

Take a look at Figure 4.3. The left is a bellows fold with alternate mountains and valleys. Cut it at the center with a plane and mirror one-half by the plane. The mirrored half section matches the remaining half with no gap, as in the center illustration. The right illustration views this operation from the top. The shape after this operation can also be made just by folding one sheet.

Similarly for cones, the foldback operation can be represented by mirroring by a plane crossing the solid. Notice the change in the crease pattern after this operation (Figure 4.4)—new fold lines are added, and mountains and valleys are reversed on

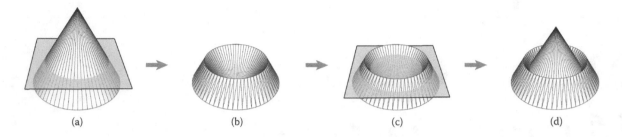

Figure 4.1 Mirror inversion operation on cone.

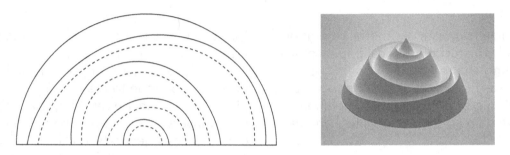

Figure 4.2 Three-dimensional origami made by mirror inversions on a cone.

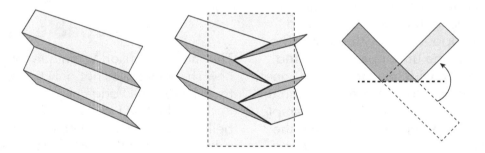

Figure 4.3 Mirror-inversion fold operation.

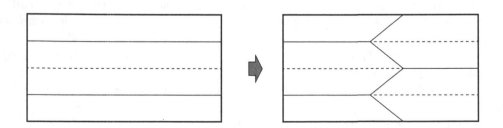

Figure 4.4 Change in crease pattern by mirror inversion.

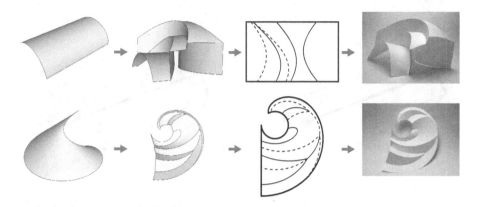

Figure 4.5   Application of mirror inversion to cylindrical and tangent surfaces.

the mirrored half. These are the only impact on the crease pattern. If the original shape can be made with one sheet, then so can the mirrored shape.

Examples in Figure 4.5 use a cylindrical surface and a tangent surface (touching a helix) as the first shape, respectively. Mirror inversion on a developable surface several times produces a variety of shapes containing curved folding.*

Note that all the curves made here are planar curves (curves resting on a plane) and not spatial curves.

4.3 Specifying Mirror Planes by a Polygonal Line

We have seen so far that 3D folding is achieved by repeatedly applying mirror inversion to a one-sheet shape. This section shows how mirror inversion is applied by specifying a polygonal line.

For handiness, the original shape is limited to a "cylindrical surface." The cylindrical

surface here is obtained by sweeping one polygonal or curved line in one direction and consists of a set of parallel straight lines. In Figures 4.6 and 4.7, the bold solid polygonal line is called the *section line* and the arrowed sweep direction is called the *sweep locus*. The sweep locus is perpendicular to the plane where the section line rests. The cylindrical surface thus obtained consists of a set of rectangles and can be made by folding or bending a sheet. Now, apply mirror inversion on this shape to make a new shape. Hereafter, the shape before the folding operation as in Figure 4.6 is called an *initial developable surface*.

Deciding the mirror plane position for this initial developable surface determines the shape obtained by mirror inversion. Several folding operations are achieved by placing multiple mirror planes.

When a part of a solid is flipped by a mirror plane, the sweep locus is flipped together. Then, the sweep locus travels as it bounces off the mirror plane, as in Figure 4.7a. If the plane, where the section line rests, and mirror surfaces are placed vertically, the sweep locus keeps traveling horizontally. It goes like a light beam reflecting from a mirror at the same incident and reflection angles, as Figure 4.7b shows.

---

* At this writing, the ORI-REF software for assisting mirror inversion is available to the public at the following URL: http://mitani.cs.tsukuba.ac.jp/ori_ref/ (accessed on March 7, 2016).

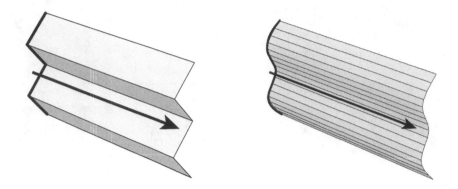

Figure 4.6   Initial developable surface swept by section line (bold solid line) along the sweep locus (arrow).

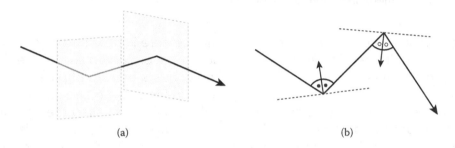

(a)   (b)

Figure 4.7   Sweep locus reflected by mirror planes.

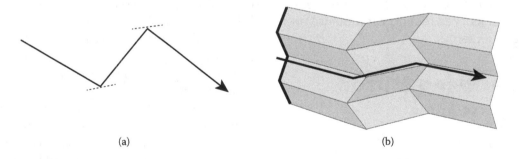

(a)   (b)

Figure 4.8   Sweep locus determines the mirror plane arrangement and thus the shape after folds.

The mirror plane arrangement determines the course of the sweep locus. Conversely, the mirror plane arrangement can be obtained from how the sweep locus travels. When designing a shape, a quick rule of thumb is to first think of the sweep locus shape, and then calculate the mirror plane arrangement. This is because it is difficult to imagine what shape you will get from the arrangement of mirror planes, but the sweep locus properly represents the shape after folds (how the straight line on the cylindrical surface zigzags).

The zigzag arrow sweep locus in Figure 4.8a determines the arrangement of the two mirror planes (thick dotted lines). Figure 4.8b is the result of applying the inversion operation on these two mirror planes to the initial developable surface swept by the thick polygonal section line of Figure 4.8b. In other words, a one-sheet solid can be determined by two polygonal lines: one represents the section line, and the other represents the sweep locus. The sweep locus is the shape viewed from the top and thus helps you design more intuitively than figuring out the mirror plane arrangement.

4.4 Relation between Sweep Locus and Shape

The folded shape can become a "closed cylinder" with the edges matched after folds under the following conditions:

1. The sweep locus start and end points match.
2. The direction of sweep locus travel matches at the start and end points.
3. The sweep locus foldback count is an even number.

The above conditions 1 and 2 are easily understandable. Condition 3 is about the inversion of mountains and valleys at every foldback. For the mountain and valley orientations to match at the edges, the foldback count needs to be an even number.

In Figure 4.9, the square and regular hexagonal sweep loci (start with a black circle) have four and six foldbacks, respectively, satisfying above conditions 1, 2, and 3. Consequently, the bellows-fold initial developable surface gives the closed cylinders as Figure 4.9b and c.

We have seen so far that a 3D shape is determined by a section line and a sweep locus. One more point is the relative positional relation between the initial developable surface and the sweep locus, as it affects the resulting solid shapes. Figure 4.10 depicts how the sweep locus (thick polygonal arrow) determines two mirror planes (thick dashed lines). The straight

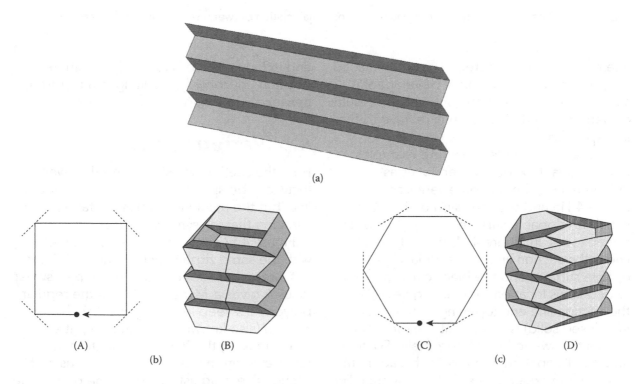

(a)

(A)   (B)   (C)   (D)

(b)   (c)

Figure 4.9   Creation of a closed-end solid.

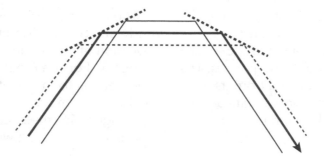

Figure 4.10   Difference in shape after mirror inversion depending on relative position to sweep locus.

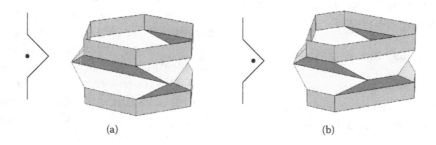

(a)   (b)

Figure 4.11   Difference in outcome depending on positional relation between section line and sweep locus.

line elements (thin dotted line and thin solid line) that make up the initial developable surface give different mirror inversion results depending on their positions relative to the sweep locus.

Figure 4.11 provides concrete examples. For a regular hexagon sweep locus as in Figure 4.9b, the outcome varies as in Figure 4.11a and b depending on how the initial developable surface is arranged for the sweep locus. In Figure 4.11, the polygonal line is the section line of the initial developable surface, and the black circle is the sweep locus location for it. In Figure 4.11a, the section line end (opening of the resulting closed solid) is located at the same position as the sweep locus horizontally. So, the final solid's opening is a regular hexagon, the same as the sweep locus. But, the section line

end in Figure 4.11b is located at a different position horizontally, resulting in a distorted hexagonal opening.

4.5 Various Shapes

Both the section line shape and the sweep locus can be specified by a simple polygonal line, but the resulting shapes are far richer in variation than axisymmetric ones.

Figure 4.12 shows example solids made with the same zigzag section line and different sweep loci. Sweeping in the zigzag sweep locus A gives a *Miura fold* and in the regular polygonal sweep locus B gives a *Yoshimura pattern* (also known as a diamond pattern). When making this Yoshimura pattern, the relative position between sweep locus and section line is adjusted so that the part of the line element folded back at the mirror plane is exactly zero length.

This very simple, two-polygonal-line input can reproduce the existing well-known patterns and even make complex shapes as in Figure 4.13. Figure 4.13 shows the section line, sweep locus, and solid shape from the left. The vertical dotted line over the section line indicates the horizontal position of the sweep locus relative to the section line.

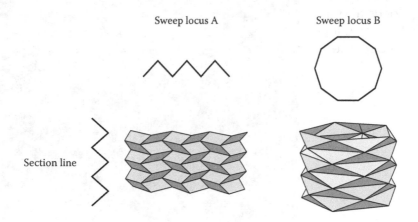

Figure 4.12   Reproduction of Miura fold pattern (left) and Yoshimura pattern (right).

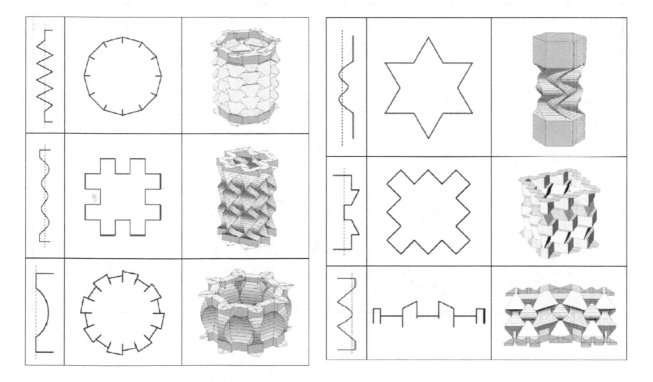

Figure 4.13   Examples of solid origami obtained from section line and sweep locus.

Hexagonal Column Sculpture

This diphycercal, sculptured column has circles in the middle of it. The points that collect six fold lines are realized by positioning the sweep locus and the section line so that the fold lines that should be there become zero length.

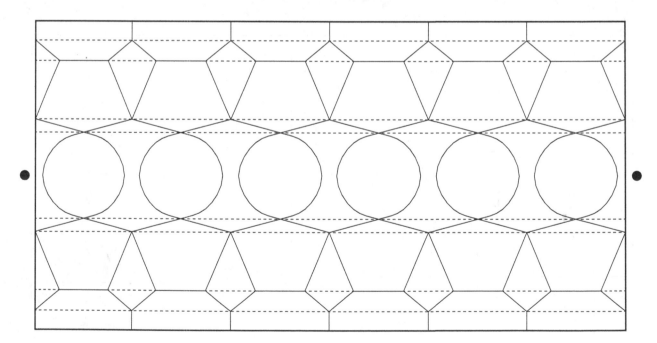

Fish-Scaled Column

This cylinder has many tiny projections. Folding takes time when working with lots of fold lines. At the top and bottom, the inner rim is a regular dodecagon with pleats tucked inward. The section line and sweep locus for this shape are provided in Figure 4.13.

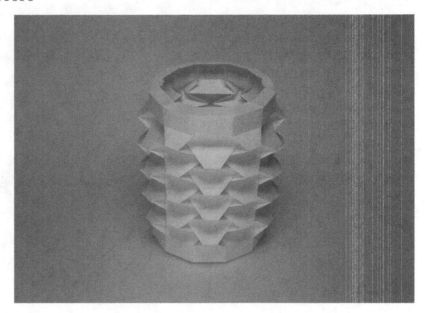

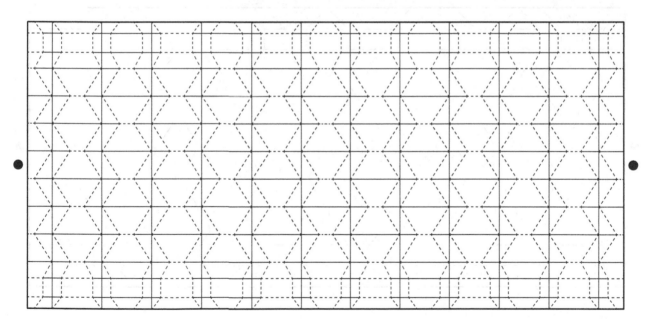

Corrugated Column

This shape has eight wavy projections in total—two each arranged on four faces. For stabilization after folds, the top and bottom consist of straight lines. The section line and sweep locus for this shape are provided in Figure 4.13. The sweep locus looks like a sharp (♯), but the opening is a cross.

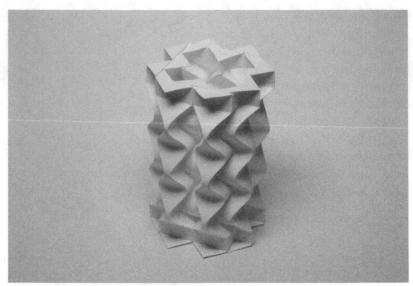

Wavy Hexagonal Column

This spring-like shape appears to be elastic at the center but is not supposed to be made so. Still, applying force from the top compresses the shape, probably due to flexibility or invisible deformation of paper. The section line and sweep locus for this shape are provided in Figure 4.13. The sweep locus is a star.

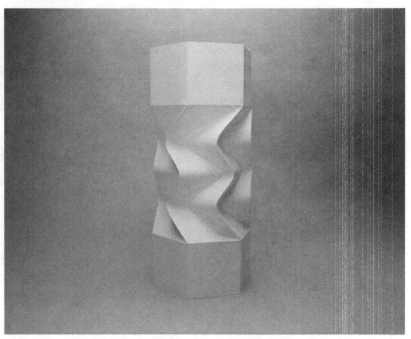

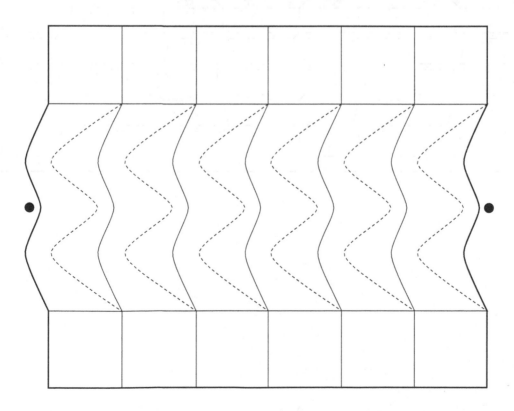

Crystal Boundary

In contrast to curved folds that express the soft aspect of paper, straight folds represent its hardness and strength. These folds make the shape stronger than a simple square pillar. The section line and sweep locus for this shape are provided in Figure 4.13.

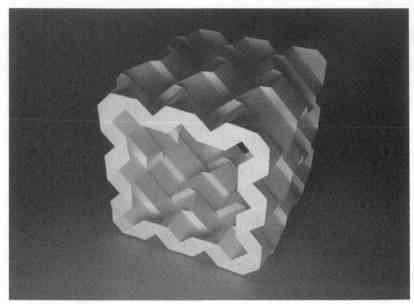

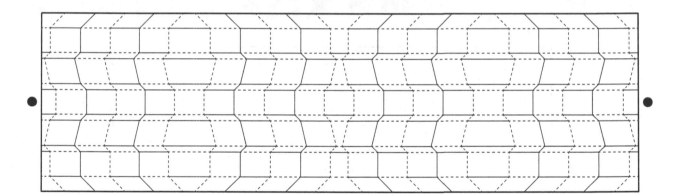

Bangle

This ring shape has curved surfaces on its outer face in the center. The crease pattern is very long because the shape is narrow in width and has many projections and recesses in the circumferential direction. The section line and sweep locus for this shape are provided in Figure 4.13.

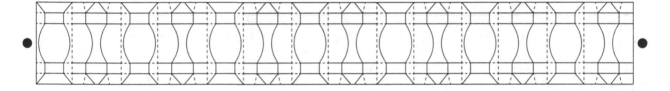

Turbine

With the regular hexagonal front face, this shape consists of a total of 18 pleats in three different sizes. The crease pattern is long for its many pleats. It would have been easier to make this with straight lines only, but I dared to also use curved lines. The outcome looks like a turbine rotating in the wind.

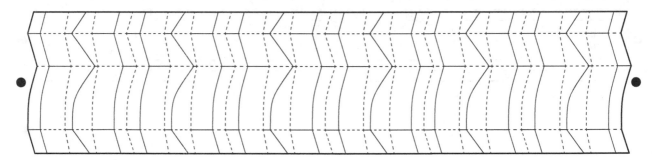

Wave Packet

Longitudinal waves in a bunch are rolling laterally.

The section line and sweep locus for this shape are provided in Figure 4.13. With the open sweep locus, the outcome is a flat, concavo-convex shape.

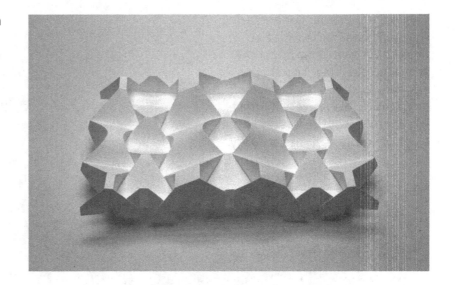

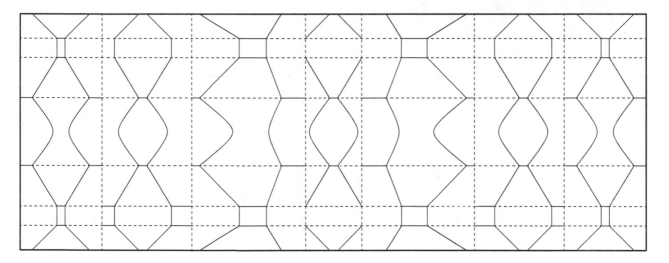

Three-Boat Container

This is a three-boat connected shape. This is made with a thick sheet as it does not show so many fold lines. The shape appears to serve as a container. Its backside also is very interesting. One fold at the left and right ends contributes to the stability.

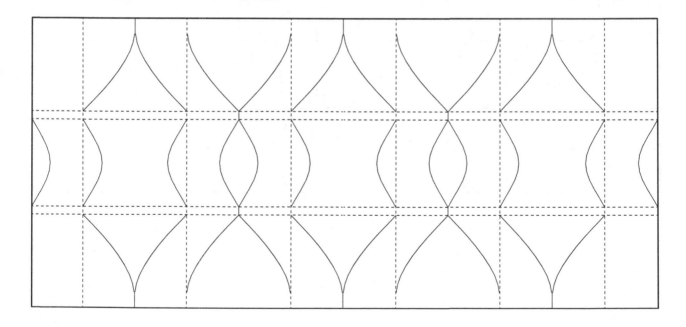

Appendix: A Big Cardboard Art Object

Once in a while, making something big is fine. We made a 2-meter-high cylinder-type origami object with six curved dents on its outer face in the center. For this height, the crease pattern had to be 1.4 meters wide and 6 meters long. To keep the shape, cardboard was selected as the material because plain paper cannot maintain the shape in this size. We worked very hard on the cardboard for folding and finally found a way to easily fold the cardboard by cutting it at the fold lines into parts, and then taping them together. This method may not sound like origami, but breaking it down into small parts simplified the packaging and delivery. The final object was taller than an adult, which had impact. Unfortunately, the object became a bit distorted under its weight. For a big object, the material weight and strength also need to be taken into consideration.

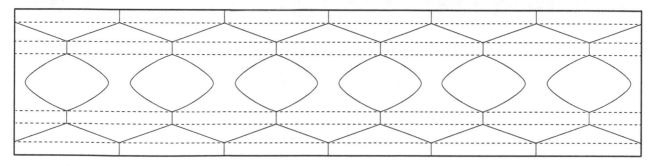

Chapter 5

Application of Mirror Inversion

Mirror inversion adds new "folds" on a solid. Devising the mirror plane arrangement makes it possible to design solids with projections and dents engaging with each other and brings it one step further in sophistication.

5.1 Curved Fold Units Combined Together

Imagine a square pillar with a square cross section. These square pillars can be placed together with no space, as in Figure 5.1.

Then, what about a square pillar partly projected and recessed? It can also be placed tightly together if its projection engages in another's recess when turned 90 degrees. Figure 5.2 (right) shows four such identical square pillars arranged two by two, side by side. In this way, as many of the same-shaped solids can be combined as you desire. The solid in Figure 5.2 with an intriguing property like this can be made from the crease pattern shown in Figure 5.3. How is a solid like this designed? This can also be made through the mirror inversion operation explained in Chapter 4.

As shown in Figure 5.4 left, the square sweep locus (dotted arrow) determines the mirror plane placed at four corners.

Mirror inversion reverses the projection and the recess. The projection fits in exactly with the recess, as they are identical in amount with respect to the original sweep locus (because it is a mirror image). The original sweep locus is a closed square. Consequently, the shape can be arranged on the grid alternately at 90 degrees to one another to engage its projection in another's recess, as shown in Figure 5.4 right.

We have already seen that mirror inversion can transparently produce a shape with projections and recesses that are exactly engaged with each other. Now, all we have to do is define the sweep locus so that the conditions are satisfied as described in Section 4.4 for making a closed cylinder solid and arranging a solid on a plane.

The sweep locus can be cruciform, as in Figure 5.5. It can be combined one after another at its projection and recess. Figure 5.6 is an example of a solid created using this sweep locus. The section line in Figure 5.6 (left) and the cruciform sweep locus in Figure 5.5 produce the solid in Figure 5.6 (right). The photo and crease pattern for this solid are provided on page 75. The solid's flat top and bottom are regarded as projections and recesses in zero thickness. Their folded portions can be tucked into engagement.

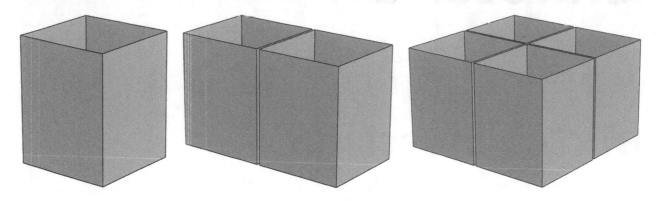

Figure 5.1   Square pillar arrangements.

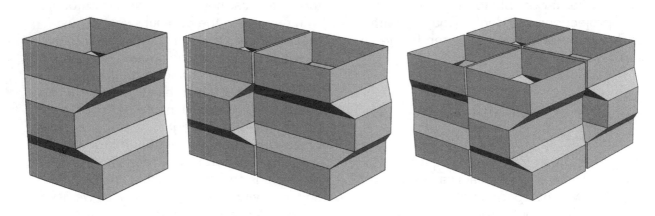

Figure 5.2   Concavo-convex square pillar arrangements.

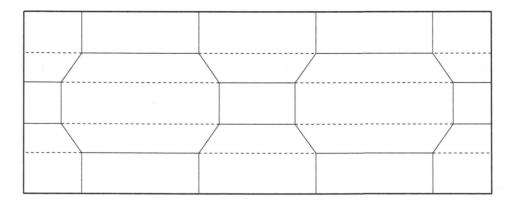

Figure 5.3   Crease pattern of solid shown in Figure 5.2.

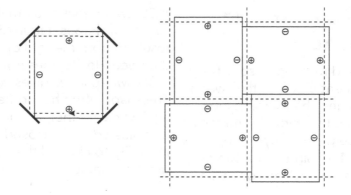

Figure 5.4   Projections and recesses generated by square sweep locus.

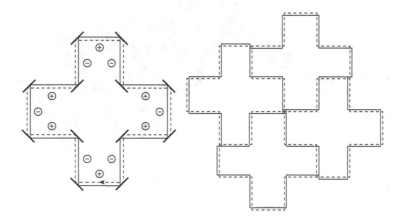

Figure 5.5   Projections and recesses generated by cruciform sweep locus.

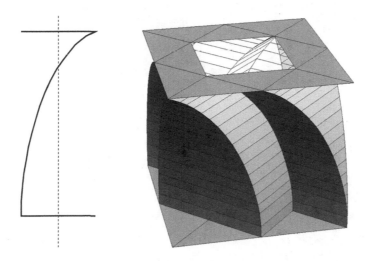

Figure 5.6   Section line (left), together with solid (right) generated from its cross section using the sweep locus in Figure 5.5.

5.2 Inversion by Oblique Mirror Plane

The sweep locus–based method of specifying mirror plane arrangement assumes the sweep locus to be horizontal and the mirror plane to be vertical. However, the mirror plane for mirror inversion can be placed anywhere, as described in Chapter 4, posing no problem for the sweep locus to be a 3D polygonal line.

A 3D sweep locus can make a more complex variety of shapes. Figure 5.7 is an example resulting from the fold operation in which the sweep locus does not rest on a plane. A 3D sweep locus gives you more freedom. The shape often has some of its sections hit against each other, and it is difficult for it to be adjusted into a closed-end cylinder. Still, its ability to expand the design range is a great advantage.

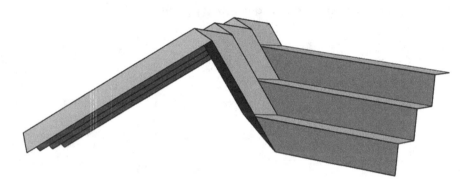

Figure 5.7   Example of mirror inversion where sweep locus is not on the plane.

Let's Make It (5) Corrugated Square Pillar Unit

This combinable, wavy square pillar fits exactly into another when turned 90 degrees at the projection and recess. See page 77 for a photo showing these units combined. Let's make several units and combine them.

PREPARATION

- Prepare the crease pattern on the next page photocopied (a bit enlarged) on a thick sheet.
- Precrease all the fold lines using a hard, pointed tool. For curved lines, cut out the template and trace along its edge to precrease neatly.
- The printed side will be the inside of the solid.
- A "glue tab" is provided at the edge for closing cylindrically.

PATTERN SHEET

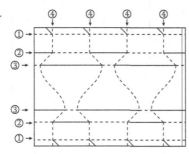

FINISHED PHOTO

PROCEDURE

1. Fold the two lines ① inward. Fold the two lines ② in both orientations (mountain and valley) so that you can fold them in either direction later.

2. Fold the two lines ③ to the opposite side of ①. You will have to fold some of them reversely, but do not mind at this step.

3. Valley-fold all four lines ④, three dimensionally as you wave the working faces.

4. Make sure the entire workpiece looks like the finished photo. Then, glue both ends together.

5. Fix the shape. Wait until the glue is completely dry.

6. Fold the top and bottom openings inward. Glue the four corners to fix. Done.

"Corrugated Square Pillar Unit" Crease Pattern

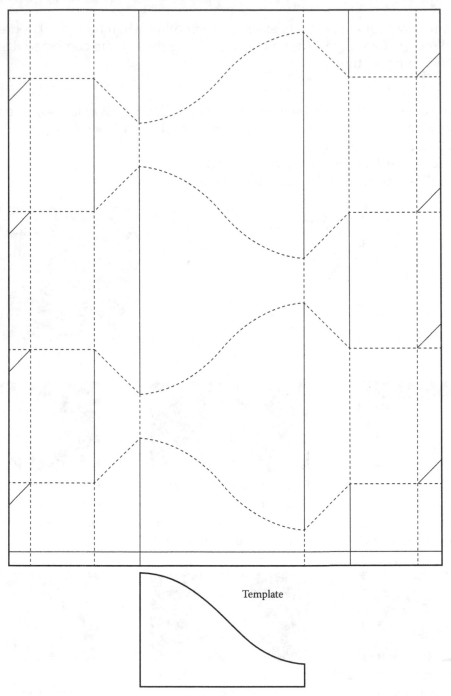

Template

The handy size is that the crease pattern's width fits that of the A4 sheet in portrait. This crease pattern is downloadable from my webpage http://mitani.cs.tsukuba.ac.jp/book/3d_origami_art/.

Combination of Cubic Units with Curved Folds

This 3D origami piece is presented in Figure 5.6. The sweep locus is a cross, but the top and bottom are a square. One unit is combined to another with the flat folded portions inserted alternately into the counterpart, making the combined units stable. At the center in the photo, the six-unit combination has two twist-closing structures where four units overlap each other at the edge.

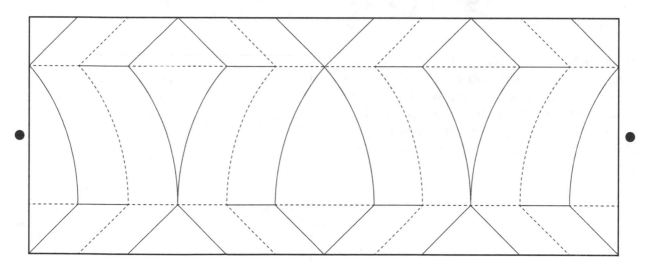

Corrugated Hexagonal Column into Honeycomb Structure

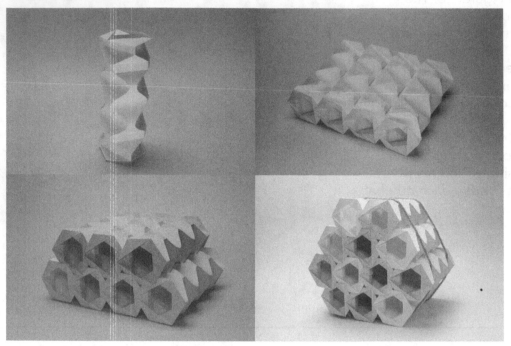

This regular hexagon–based solid can be stacked like a honeycomb structure. Whether the unit faces forward or backward, it can fit exactly with another at the projection and recess due to its upper–lower symmetry. Three units overlap each other at the edge to make a twist-closing structure.

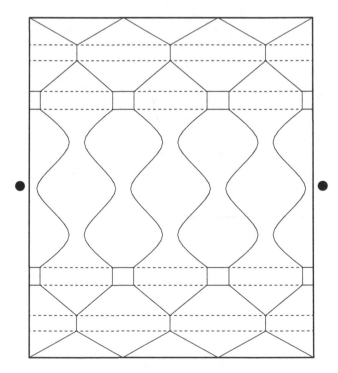

Combination of Corrugated Square Pillar Units

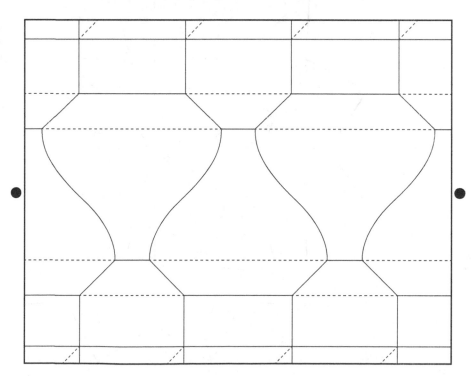

This shape is presented in "Let's Make It" on page 73. Both ends are a square. The edges can be combined with no space. In the photo, the units have identical shapes but are in different colors. Combining the shapes made from different color sheets also adds an amusing touch.

Stellar Medallion

Mirror inversion is applied several times to a star-shaped, one-sheet pyramid without cut. This is the same idea as flipping the cone presented on page 51. The tilt of the mirror planes is adjusted so that multiple fold lines can intersect at a point.

Patchwork of Curved Surfaces

Mirror inversion is applied several times to a simple sloping cylindrical surface. The smoothness of curved surfaces provides a pleasing contrast to sharpness of fold lines and produces a beautifully shaded shape. I came up with this shape through unintentional trial and error.

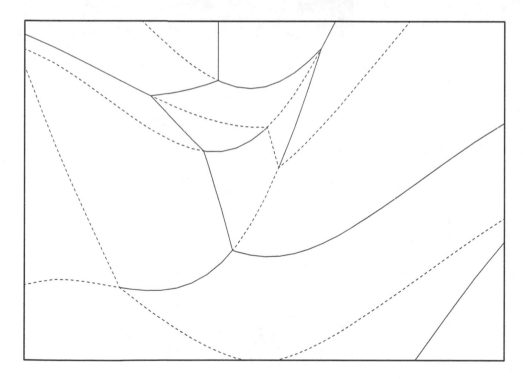

Wave Relief

This shape is created through several mirror inversions to a two-humped wavy cylindrical surface. This seems simply done, but creating a one-sheet shape like this is a tough thing without the help of a computer.

Kabuki Face

The curved folds used in this work are obtained by oblique mirror planes. Interestingly, all the central fold lines are mountains. Though the crease pattern is very simple, it took a lot of effort to design it so that the left and right edges matched to close the structure. This is one of the most difficult design categories in contrast to ease of making.

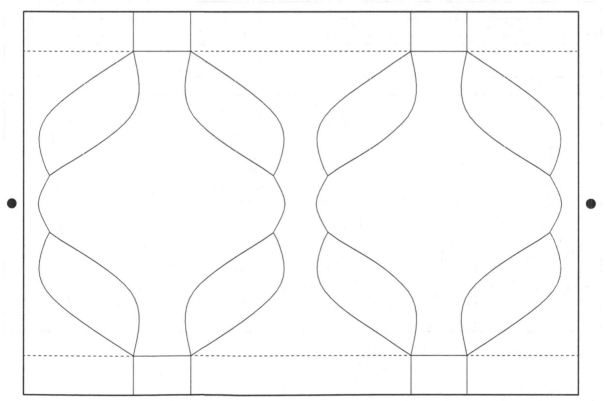

Star Pottery

The simple, 3D-pleat cylinder–type lower portion is connected to the intricately terraced upper portion obtained through mirror inversion. Combining solids made by different methods widens the variation. The bottom is closed by tucking in the pleats without gluing.

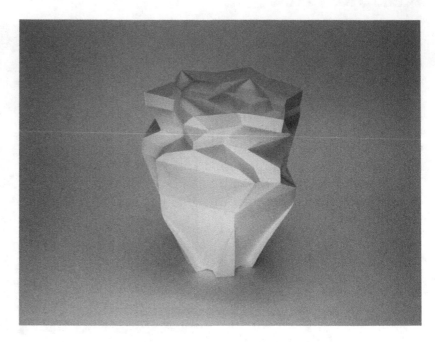

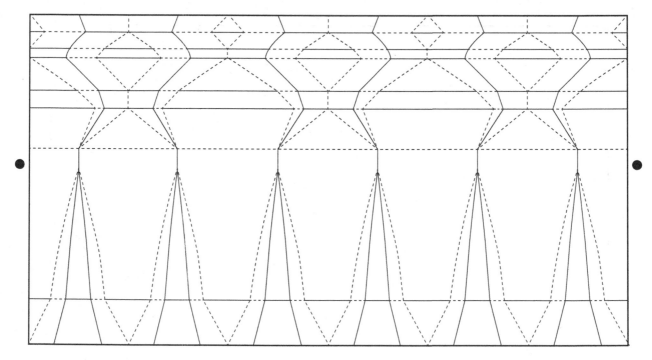

Tulip-Top Milk Carton

With a square pillar as the base, the upper half is made into a tulip flower shape. An oblique mirror plane is used. The shape looks down at the center. Though the shape is simple, it is difficult to make the shape closable like a cylinder and the tulip portion's bottom connectable to the square pillar. The top end is a little distorted but can be opened like a milk carton.

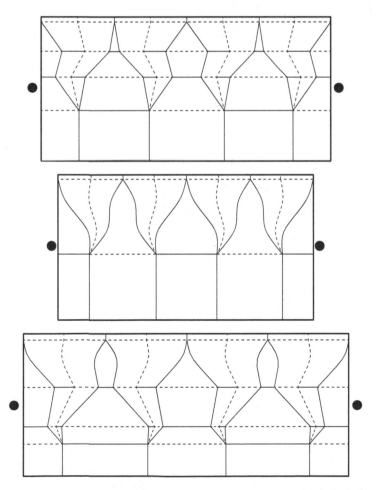

Chapter 6

Voronoi Origami

Three-dimensional origami design is a really fun way to make planar materials into a solid—in other words, transforming two dimensions into three. There is another kind of pleasure in origami forms folded flatwise. Particularly, the tessellation fold structure tiled up with polygons is stunning with a light behind it, presenting a geometric mosaic silhouette.

6.1 Tiling with Different Polygons

Regular triangles, squares, and hexagons can be tiled on a plane. We have seen on page XIII that the tessellation pattern is made by this polygon-based twist folding.

It is known that for a tessellation pattern, the crease pattern can be created by a simple procedure of "shrink-and-rotate polygons" as in Figure 6.1.

First, shrink each regular polygon tiled on a plane in equal proportion with its centroid as the center. This produces a clearance in between the polygons. Then, connect the vertices that once were at the same position to make a new polygon. If the number of vertices at the same position is four, the newly created shape is a square, as in the figure. Now, rotate the shrunk polygons by the same

angle together with the new polygon. The pattern thus obtained can be folded flatwise because each vertex satisfies the local flat-folding conditions on page X.

It is known that the tessellation crease pattern can be created from a tiling of different regular polygons in the same way above. Out of the tiling patterns that cover the plane by regular polygons, there are only eight Archimedean tilings for which all vertices are uniform in shape (on page 42). If the vertex shape is allowed to be nonuniform, then there are infinite numbers of tiling patterns. Therefore, this simple shrink-and-rotate operation can generate countless numbers of tessellation patterns.

Tess, software capable of generating tessellation crease patterns by the above method, was developed by Alex Bateman and is available on the Internet.[*]

Figure 6.2 shows the tessellation pattern generated in Tess based on a regular triangle and hexagon tiling. The center diagram is the crease pattern obtained by the shrink-and-rotate operation on polygons, and the right diagram is the silhouette after folds.

---

* "Tess: Origami tessellation software." http://www.papermosaics.co.uk/software.html (accessed on February 12, 2016).

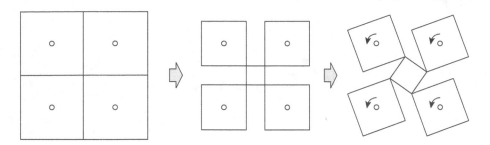

Figure 6.1 Shrink-and-rotate operation.

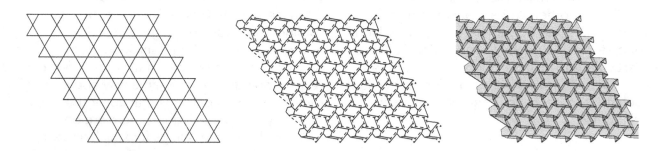

Figure 6.2 Tessellation pattern generated in Tess based on regular triangle and hexagon tiling.

6.2 Origami by Voronoi Tiling

Can the shrink-and-rotate operation make a tessellation pattern from a tiling other than regular polygons? As a matter of fact, this method is very powerful. Robert Lang and Alex Bateman have proven that the combination of any shaped tile is flat-foldable if it satisfies the condition that "the line connecting the two adjacent tiles' rotation center makes 90 degrees with the side where these two tiles contact."

Figure 6.3 presents a *Voronoi diagram*— a plane partitioned into cells in terms of "which point is the closest" based on the random points placed first (called *generators*). In another expression, a Voronoi diagram is obtained by gradually widening the domain starting from each point

at a given speed and setting a border at the position where two domains meet. In a Voronoi diagram, the line segment is located at an equal distance from the two nearest generators (the perpendicular bisector of the line segment connecting two points). Therefore, for Voronoi diagrams, the flat-foldable crease pattern is obtained through the reduce-and-rotate operation with the generator as the rotation center.

Figure 6.4 shows the fold line pattern obtained from the Voronoi diagram in Figure 6.3, by shrinking each region 30% and rotating 20 degrees. The mountain and valley assignment is underspecified. The pattern can be folded flatwise in various ways.

The shape from a Voronoi diagram is determined only by the generator location. Giving a rule to the generator arrangement will make a beautiful symmetrical shape.

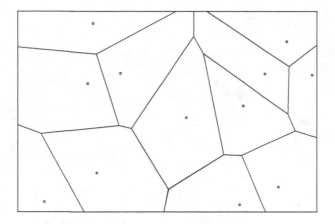

Figure 6.3   Voronoi diagram obtained from points randomly placed on a rectangle region.

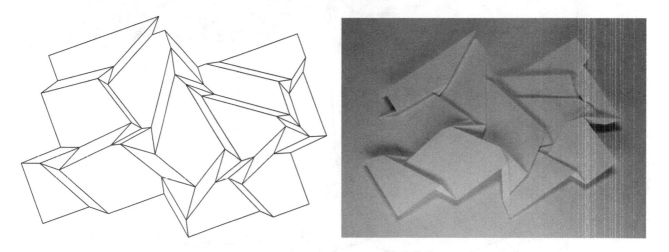

Figure 6.4   Fold line pattern from Figure 6.3 Voronoi diagram, together with the folded shape.

Star Voronoi Origami

This is a Voronoi diagram-based star shape. The Voronoi diagram generators are arranged so as to have five rotational symmetries. Folded down flatwise, this origami is silhouetted beautifully by a light behind it.

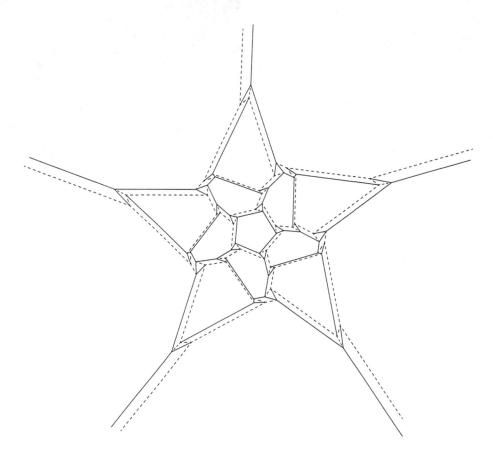

Voronoi Origami Crest

This shape is based on the seldom-used septagon. From the center, septagon, small pentagon, and large pentagon tiles are arranged. All the fold lines work together during the making, involving a different sort of difficulty from 3D origami.

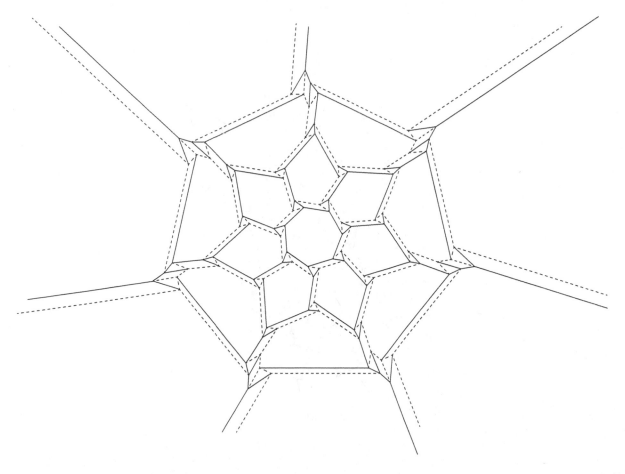

Network of Polygons

This complex pattern is created by rotationally symmetric shapes arranged onto rotationally symmetric places. The pattern looks random but actually has double rotation symmetry. The crease pattern is partially cut out, because folding the whole body is quite tough.

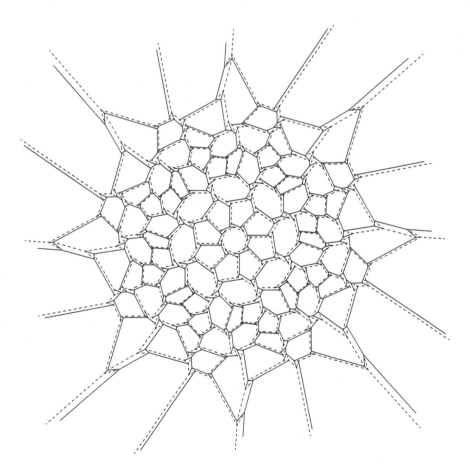

Medal

This was created from a Voronoi diagram having six-time rotational symmetry. The shape is finalized by turning the central regular hexagon. This form has a sharp impression. The original generators are carefully arranged so that the set of triangles makes up the shape.

Six-Blade Pinwheel

With some additional curved lines, the "Crest" on page 91 is shaped into three dimensions. The right-side version uses a checkerboard sheet to depict the rotating center part. The curved lines are determined not from calculation but are fine-tuned through my senses. This shape holds because a few mismatches are absorbed in the paper with its minute elasticity and distortion. Enjoy freewheeling thinking to make new shapes.

Appendix: Shapes Folded Out of Lattice Patterns

The lattice pattern below can be easily folded from a square sheet.

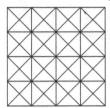

Using this lattice pattern, popular shapes can be made such as a trick boat and a pinwheel.

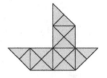 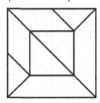  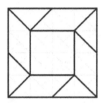

Then, how many shapes can you get from this lattice pattern? It is a total of 13,451 shapes found from my computational research in collaboration with Yohei Yamamoto. A simple lattice pattern can make a surprising number of shapes. The crease patterns below are some we picked up that resemble alphanumeric characters after folds.

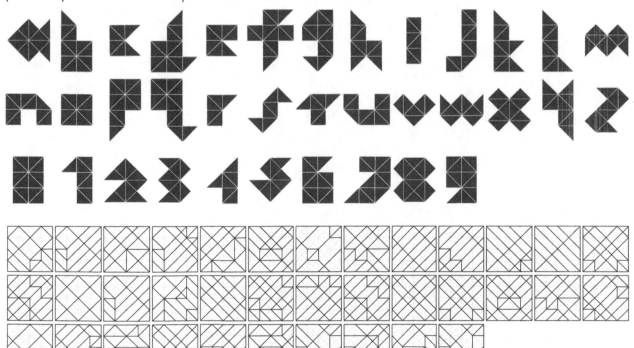

Chapter 7

Various Origami Designs

So far I have explained 3D origami design methods divided into several categories. You now know that one design method can bring forth a variety of shapes based on it. Furthermore, you can get more variations by connecting multiple 3D origami shapes made by different methods.

This chapter introduces you to more advanced origami creations—some are made by a combination of several design methods and others in fascinating shapes extended beyond the design methods explained so far. Each of them is quite unique, with an individual theoretical background or special cut-ins.

Diamond-Cut Hat

This diamond-cut shape presents brilliant shades made by radially arranged triangles. Stability is achieved by the tucks inside at the portions away from the center. The result is this hat-like shape.

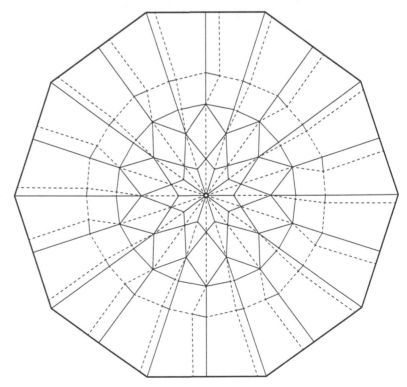

Bavarian Mold

This is an example of a torus shape realized by "cutting," a prohibited technique in origami. The cut-ins are shown by the bold lines at the center of the crease pattern. At the cut-in, the pleated place switches from outside to inside. The resultant shape looks like a "Bavarian mold" for cooking Bavarian cream.

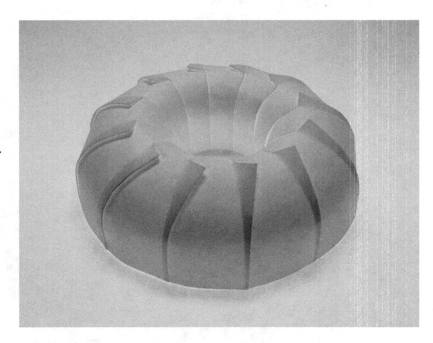

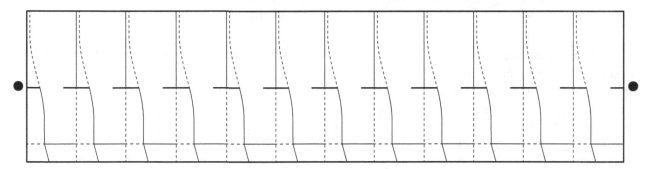

Accordion Capsule

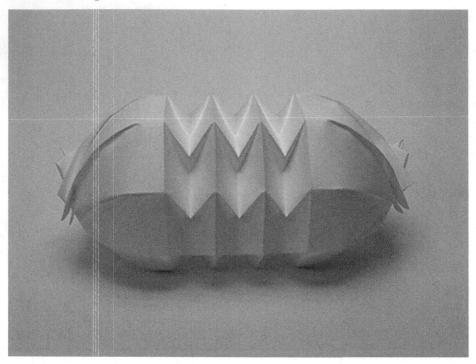

This solid has a bellows at the center and a hemisphere on the right and left sides. Pushing or pulling it on both sides compresses or expands the bellows, though the structure is not supposed to deform in design. Still, the physical creation can compress or expand due to distortion of paper or movement of fold lines.

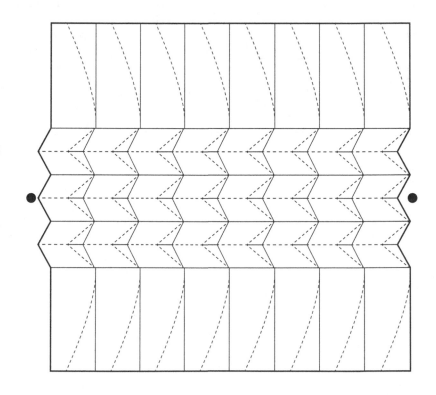

Whirl

This spiral pattern is made with combined concentric rings of curved lines. The shape is flat as is but becomes an orderly concavo-convex by folding back the two portions between the straight mountain and valley fold lines on the crease pattern.

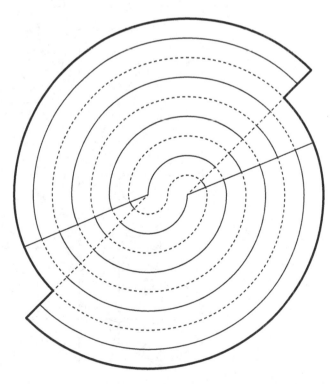

Star Tessellation

An example of tessellation tiled with geometric 3D structures on a plane. Its star-shaped tiles are based on a hexagon. Folding takes time working with lots of fold lines. Of course, as many solids can be connected as you desire. But, it is hard work to neatly fold sections away from the paper edge.

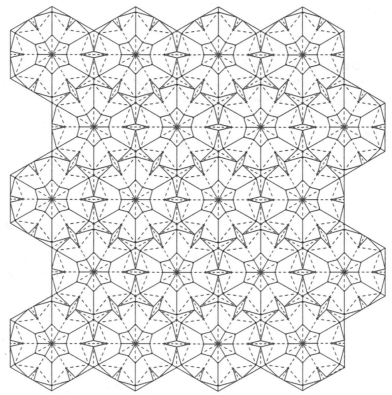

Chapter 8

Conclusion

By now you have learned much about origami's wonderful ability to create beautiful geometric shapes, sometimes with curved surfaces, just from the simple operation of "folding a sheet." This is achieved through close observation of the relation between the resulting fold lines and the folded shape.

Origami research is conducted not only for making "shapes," but also for various points of view in many fields such as mathematics, engineering, education, and biology. I would like to outline this origami-related research in personally defined themes.

8.1 Origami Design Techniques

An origami creation may be born from folding a sheet until it turns into a shape that looks like something. In contrast to this well-practiced "pareidolic" approach, "origami design" is preplanning the fold line arrangement to fold the desired shape with a single sheet of paper. Jun Maekawa's "Devil" (1980), which was carefully worked out in detail up to fingertips, is recognized to have opened the era of origami design. Origami design is one of the themes that intrigues me most. In origami design, you have to consider what condition must be satisfied for a single sheet to turn it into a shape, requiring deep insight of origami geometry.

In the past, a wide range of research was conducted on shapes resulting from flat foldings. The tree-figure approach was invented and researched by Toshiyuki Meguro and Robert J. Lang. This approach designs a crease pattern by using a skeleton structure, one that looks like a tree, to express the desired shape. *Origami Design Secrets: Mathematical Methods for an Ancient Art** systematically presents this approach and others including the box pleating design method (combination of rectangular folding areas).

8.2 Rigid Origami

Rigid origami is a foldable "structure made of a polygonal rigid material connected with hinges." Twist-folded structures are not rigid origami, as these are folded away by being twisted. Most origami creations are not rigid, as they contain a folding process making use of paper's property of elastic deformation.

The Miura-ori structure in Figure 8.1 is rigid origami with a single degree of freedom and can be folded and deployed by holding onto one section and pinching a different section to move the whole body. A structure like this is a key feature in making industrial products and architectural structures. Forming an intended shape by rigid origami is a very important and practical theme. There are many research themes

---

* Robert J. Lang, *Origami Design Secrets: Mathematical Methods for an Ancient Art* (Boca Raton, FL: CRC Press, 2011).

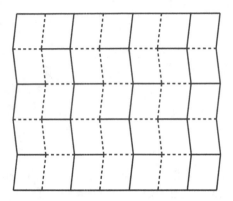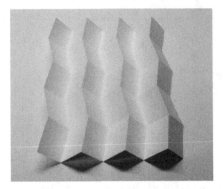

Figure 8.1 Miura-ori pattern.

related to rigid origami, such as shape designing, judgment of rigid foldability for given crease patterns, and simulation of the folding and deploying process. In recent years, rigid origami has been studied intensively.

8.3 Curved Folds and Curved Origami

Many of the creations presented in this book have curved surfaces. Folding a sheet along curves creates a shape containing curved surfaces. Lots of curved origami works were created by mathematician David Huffman. His design technique makes wide use of mirror inversion with developable surfaces, as described in this book. But the fold lines created by mirror inversion are limited to planar curves. Designing shapes containing space curves is still a challenging problem. As a solution, there is an approach discretizing the shape by a set of square planes. It is known that if concentrically arranged fold lines are mountain- and valley-folded alternately, then an intricately curved solid is formed due to balance with the stress generated on the material. It has not yet been revealed how this shape is represented mathematically. There is still much room for further research into space curve folds.

8.4 Computational Origami

Computer-assisted attempts to solve origami problems are called *computational origami*. This field has become popular since the 1990s, when origami design techniques were systematized. In recent years, computers have been utilized in various fields of origami research such as pattern enumeration, fold simulation, and interactive origami design.

The use of computers makes for more efficient and precise origami design and simulation and advances the development of new fields. It contributes to the creation of manually impossible designs. Origami research will be further promoted by combining origami mathematics, algorithms, data structures, and user interfaces.

8.5 Origami with Thick Materials

For the origami techniques to apply to industrial products, material thickness is a crucial feature that cannot be ignored. Strictly speaking, a thick material cannot be folded flatwise. The 2D plane problem becomes a 3D problem. Interference occurs at intersections where fold lines cross due to thickness, requiring some creative thinking to treat intersections. In some cases, you have to consider material elasticity and the occurrence of distortion and wrinkles at folded/bended portions. Proper modeling of a form folded with a thick material is a problem far more difficult than ideal origami that assumes zero thickness. There is a broad range of challenges, including the fold-away mechanism and the handling of place of collision.

8.6 Robots and Origami

A robot-automated folding process is desired for the efficient manufacturing of paper-folded products. Robotic paper folding is a difficult field fraught with a great many challenges. There is research on robotic folding operations using a robot with a suction cup and a thin plate or a robot arm with several mechanical linkages. Some manufacturers have their manipulators do origami to demonstrate dexterity. Robots will not be able to fold a crane by themselves anytime soon.

Much research is being conducted on the incorporation of origami know-how into robot mechanisms. *Oribotics* is a new term derived from it. A science fiction–like world is coming where a robot made of a flat material transforms like origami and goes into action.

8.7 Relation between Living Things and Origami

When an insect emerges, its crumpled wings are expanded into surprisingly large dimensions. The same goes for flower buds when they bloom. Origami knowledge is useful in understanding these mechanisms. Conversely, learning from them helps origami research—how living things store a big membrane in a small space and how they open it in a moment. For instance, the way a dragonfly opens its wings at emergence is

applied to the reeling and extending of large membranes. Some scientists are working to apply this *biomimicry*, or learning from living things, to the field of origami.

8.8 Origami and Mathematics

Paper folding is closely related to geometry. Origami research has been conducted in the field of mathematics since olden times. Mathematical knowledge, such as conditions for mountains and valleys to be folded flatwise (Maekawa's theorem and Kawasaki's theorem), is fully utilized in origami design. In addition to design, there is an origami-based approach to math problems—solving a cubic equation or trisecting an angle through folding operations, folding a regular polygon through the basic folding operations (known to fold up to decagon), and more. Determination of origami-related complexity of calculation is another active field of research. For instance, how much computational effort would it take to solve a given problem like judging whether given fold lines are flat-foldable and enumerating sheet overlap order? The book by Demaine and O'Rourke, *Geometric Folding Algorithms: Linkages, Origami, Polyhedra*[*] covers these topics.

8.9 Origami and Education

There are attempts to introduce origami for use in elementary mathematics education. This includes the bisection of angles, triangle's properties, and so on. For instance, the crane crease pattern can be a geometry learning tool for right triangles,

---

[*] Erik D. Demaine and Joseph O'Rourke, *Geometric Folding Algorithms: Linkages, Origami, Polyhedra* (New York: Cambridge University Press, 2007).

angle bisector, and triangle incenter. Hands-on experience increases learner interest and concentration. Appropriate learning themes from origami can effectively contribute to education. Searching with "origami" and "mathematics" as keywords gives a lot of hints for education books.

8.10 Application of Origami to Industry

"Folding" technique has greatly contributed to industries whose materials are not limited to paper. The most commonly cited examples are foldable satellite solar panels and embosses on beverage cans. Some familiar examples of the application of origami technique are candy boxes, paper bags, folding fans, and umbrellas. Recently, a foldable kayak has been introduced. However, not so many new industrial products have emerged that employ the latest folding techniques. One of the reasons for this is that "making by folding" prevents mass production. If a container is a shape in need of complex folding, it is more efficient to make it with molten resin poured into a mold. Folding a piece of material has the advantage of airtightness. If it does not matter much, you do not have to stick to making by folding a piece of material. For industrial uses, the functionality–cost balance is significant. Origami-inspired product design must take the production process into account. This is another challenging problem.

Other than folding down to a small size, "folding" is often used for adjusting material strength. Embosses on beverage cans increase strength. Research is ongoing for practical uses like partially folded vehicle side panels (for efficient collision energy absorption) and lighter weight sheet panels.

8.11 Others

There are many other areas of research that are uncategorizable, including prediction of folded shape and folding procedures from crease patterns opened after folds, how to generate folding instruction animation, origami design techniques in particular fields such as unit-folding and tessellation, and medical applications like stent folding. *Origami* has various definitions. Any research including "folding a flat thing" may be origami research. Origami is a rare, interdisciplinary research subject.

Afterword

My study and research on origami started in early 2005 based on a love of paper-crafting. At first I felt really tight with origami's simple operation "just folding," which limits the resulting shapes. That was when I learned about the Japan Origami Academic Society. Through gathering with people interested in origami and seeing works by many origami artists, I was greatly surprised by origami's power of expression and many years of research results.

I thought, "Wow it's wonderful."

Already specializing in computer-aided designing of 3D shapes, I decided to create a software program capable of designing origami shapes.

With the use of a computer, you can make various shapes freely, but remember that freewheeling is not possible when designing an origami shape. We have to follow the strict rule of "making just by folding a single sheet of paper."

So, I worked out the origami design methods presented in this book and incorporated them in a program running on a computer. As explained so far, the same basic design method can produce a totally different shape by an operation like changing the section line. Through a lot of trial and error using a computer, I discovered many interesting shapes somewhat different from traditional origami. Some of them contain curved folds and smooth, round surfaces, giving quite a different impression from traditional linear origami models. I have published the crease patterns of those creations in two books in Japan. Fortunately, the books were well received by many people as "origami that is a bit different" from traditional models like cranes or samurai warrior helmets.

This book reveals how these origami works are created. Though details are provided in papers that I have published since 2005, I have long wanted to put them in a more readable form. I am so pleased to see my long-cherished desire come true with this book.

Paper-crafting has been my life since I was a little child, and computers as well. I have been into programming my own games since I was in elementary school. Now grown up, I have again gotten hooked on personal computers and paper-crafting. The work object is replaced with "origami" without cutting or pasting, and programming is for origami design. Origami is a simple and easy pastime for everyone with just a sheet. But its deep world still leaves much room for further research. I truly feel lucky that I have something to enjoy and pour my enthusiasm into.

DOWNLOADING OF CREASE PATTERNS

The crease pattern data of origami creations in this book is downloadable from the following URL:

http://mitani.cs.tsukuba.ac.jp/book/3d_origami_art/

The crease pattern data are available in DXF, PDF, and SVG formats.

Index